A French Officer in the American Revolution

By Didier Thibault de Ménonville

Translated by
Norman Desmarais
and
Gina Doubleday

Library of Congress Cataloging-in-Publication Data

Didier Thibault de Ménonville, A French Officer in the American Revolution, a fact-based first-person narrative about a real-life French officer who fought with Lafayette and George Washington in the American Revolutionary War

Summary: This book is based on meticulous research and the journals of François-Louis de Ménonville, written by his direct descendant as if it was told by the soldier himself, with extensive footnotes for those interested in learning more. It describes not only his adventures in battle and comments on military strategy but his thoughts on George Washington, Lafayette, and other major figures.

ISBN: 978-1-939282-52-1

English language version published by Miniver Press, LLC, McLean Virginia

First edition July 2024

Editor's Note: This book is written in the first person by a descendant of François-Louis Thibault de Ménonville, an officer who fought with Lafayette in the American Revolution, based on research by the author. For the sake of historical usage, this manuscript includes some dated language that is not consistent with modern sensitivity.

For my wife, my sons, and daughters-in-law, who accompanied me and advised me along the way.

For François-Louis, without whom this book would not have been written…

TABLE OF CONTENTS

EARLY YEARS

Over the course of my life, I fought in three wars and lived through a revolution bloodier and more barbaric than all of the three. I saw the end of the world that I was born into. More recently, less occupied with my responsibilities, I have looked back on my life and felt the desire to relive these moments by recounting them. Perhaps this account will prove to be of some interest to my descendants, and for them I will briefly recall the history of our family.

Our earliest known ancestor is Armand Thibault, originally from England. He came to France with his son Jean around 1640 to escape persecution for being a Catholic. Due to his knowledge of the English language and the country, King Louis XIV made use of him on different occasions for negotiations with the Cromwell regime. His missions were led by Secretary of State Sublet des Noyers. In recognition of his service, on November 5th, 1649, Armand was appointed hearing officer, then on May 15th, 1654, Head Clerk of the Accounts Chamber of Lorraine and Barrois in Metz, and the Accounts Chamber of Nancy.

When the King dispatched him to Lorraine, he found the offices of the judiciary, police, and finance in great disarray. They had been established through simple judges' commissions without having obtained real letters of provision.

On October 12th, 1655, Armand Thibault's son Jean was put in charge of an investigation to reestablish order in these matters, and for this he received a royal pension in 1657.

Armand Thibault was from a noble family. His son Jean tried to obtain from England proof of his father's nobility, but was unable to. Nevertheless, King Louis XIV granted him nobility in 1659. On August 20th, 1733, Louis XV confirmed the decision of his ancestor and assured the Thibault de Ménonville family they would enjoy "the privileges due to the nobility issued from a noble and ancient extraction and that they could wear the same coat of arms as their ancestors."[1] The martlets, depicted without beaks or claws on our family coat of arms, signify that one of our ancestors left on a crusade and returned financially ruined.

I wish to make it clear that these letters of confirmation clearly establish that the Thibault de Ménonville are of noble and ancient extraction from their English ancestor, and not noblesse de robe due to the functions of magistrate that were carried out by Armand Thibault.

---

[1] Letter signed by King Louis XV on August 20th, 1733.

As for myself, great-great-grandson of Armand Thibault, I was born in Villé, near Baccarat, before the new chateau was built in 1774. It was a large residence flanked by four corner towers. I was sorry for the change to the chateau and the loss of the former house, which certainly dated from the XIVth or XVth century, and where I spent my childhood.

I spent my early years in Villé, and in 1755, at the age of 15, I was accepted to the school of gentlemen cadets of the King of Poland in Lunéville. To be admitted required proof of five quartiers of nobility (five noble ancestors), to be at least 15 years old, and from a Catholic family. The course of study lasted three years. The students were divided into four brigades, two made up of Poles and Lithuanians, two of young men from Lorrains and Barrisiens. We were lodged in the living quarters of the school where we each had our own room. Meals were served in a hall by a caterer from the town. Also attached to the school were a chaplain, a wigmaker, a nurse, four laundresses, and twenty-three domestics wearing the livery of the King of Poland; it was the king who covered all of the expenses of the school.

The château de Villé before its reconstruction in 1784.

I was very proud to receive the two full uniforms.

The first was a uniform of yellow cloth (suit, jacket, and breeches), trimmed with brandebourg knots embroidered in silver, the jacket with silver buttons. The second uniform was blue, trimmed in the Polish style, including the collar, with a scarlet jacket and breeches with gilded buttons. Each uniform was completed with a hat embroidered in gold and silver, and red stockings.

Our instructors were a master at arms, a dance master, and a teacher of history, mathematics, and languages. Two squires of the king led our exercises in equitation, for which twenty horses were at our disposal. Reveille was beaten on a drum at 5:30 in the morning in winter and 5:00 in the summer. We had lessons every day in languages, math, and history. I was able to learn English, which proved extremely useful throughout my career, but especially during the campaign in America. Our progress was followed closely by our instructors, especially in mathematics and history. A public rehearsal was held every two months, and those not considered worthy to pass this trial were seated separately on a bench; being excluded three times meant dismissal from the school. Discipline was extremely strict. Anything missed was punished by prison, and sometimes bread and water. It was forbidden for us to frequent taverns and game rooms, but we were sometimes permitted, during certain hours, to visit families of good reputation and go out accompanied by an officer.[2]

For me, attending this school was essential. My family did not belong to the nobility of the court, not having been presented to the king, so without this education it would have been impossible for me to progress in the army. Military schools allowed me to reach the ranks generally reserved for the high nobility, whose children entered the army very young, sometimes as young as 8 years old. Most of the time this was in the regiment of their father or a friend. They believed, accurately, that their name and their ties to the court sufficed for their advancement. However, their education level was often quite low.

After the school for cadets, where I stayed just over a year instead of the usual three, I passed the admissions exam to the École de la Fère and entered in April 1757, at 17 years old, with the rank of second lieutenant. The school had been created by King Louis XIV, who had realized that artillery was becoming essential to war, whether for open battle or a siege. Every week we had three hours of theoretical classes and five hours of practical exercises. Our main instructor, the Abbot Nollet, taught us mathematics, hydraulics, physics, and chemistry.

My performance at the final exams meant that I left the school with a rank of second lieutenant.

---

[2] Benoît, A. *L'école des Cadets-Gentilshommes du Roi de Pologne à Lunéville*. 1738–1766.

I then passed the entry examination for the École royale du génie de Mézières,[3] where I began on January 1st, 1759. This exam was difficult, as two candidates of every three were refused, and only fifteen candidates were admitted each year.[4] We lived in the magnificent buildings of the Governor's Palace.[5] The school took care of everything, providing materials and lodging. We formed a community of students, instructors, and military men. The staff sought the intellectual success of its students, the foundation of the school's prestige, and for that had established strict discipline, ceremonies, exams, and a strictly regulated social life. There were more officers and instructors than students, so we were always accompanied and supervised.

We were in the classroom or workshop for three hours in the morning and three in the afternoon, six days a week, supplemented with workshops in advanced drawing and tours of the military forts with the officers. Our work was evaluated very frequently and the exam results were sent to the Secretary of State for War, along with a briefing from the commander describing the morals, diligence, behavior, character, and scientific knowledge of each student.[6]

On the 1st of January 1760, at the age of 20, I was formally recognized as an engineer. I had earned the most prestigious diploma of the French Army. Between 1760 and 1762, I was sent to the coast of the Mediterranean, then to the island of Minorca. The island had been taken back from the English, but since then there was little activity in the Mediterranean as combat was mostly in Prussia and America. The treaty of February 10th, 1763, ending the Seven Years' War, gave the island of Minorca to England in exchange for the island of Yeu.

---

[3] The current École Polytechnique considers the École de Mézières its predecessor.

[4] *La jaune et la rouge*, March 2020, p. 21 sqq.

[5] Today the Préfecture, or administrative seat, of Mézières.

[6] Pautet, S. (2016). *Produire une élite savante et technicienne à l'École du génie de Mézières* (Artefact 4) & Robinet A. (1949). *L'École royale du génie de Mézières.* Revue d'histoire des sciences et de leurs applications. Tome 2, n° 3.

CORSICA

It was in Corsica that I first experienced war. The history of this island is interesting: the Corsicans, fed up with the oppression of their feudal lords, requested in 1358 the protection of the Genoese, who established their domination over the Corsicans with extreme brutality. The Corsicans rose up against the occupying force, who they themselves had called, but the true revolt began in 1729. At first, only one part of the population followed the movement, which was also weakened by infighting. Genoa, trying to maintain its authority, requested assistance several times from France, who was funding them. Weary of war and faced with the nationalist wave led by Pascal Paoli, Genoa preferred to withdraw and sell its rights over Corsica to France, in signing the Treaty of Versailles on May 15th, 1768.

I admired the ingenuity of the Genoese to sell an island that did not even belong to them. For the French, the main interest in having Corsica was to keep the English from moving there, and we therefore had to conquer what we had purchased. A first attempt failed in 1768. Charles-François Dumouriez was part of the expedition. He knew the situation because he had recently offered his services to the Genoese to fight the Corsicans, and since they had refused, he then went to the leader of the Corsicans, Pascal Paoli, against the Genoese, but he had also turned him down. He therefore found himself at the service of the French. He and I crossed paths several times, first in Corsica, later in Poland, and finally during the revolution. He was intelligent and very brave, but intriguing in the extreme and constantly seeking money and honors. According to him, the failure of 1768 could have been avoided if we had only listened to him.

In 1769, Versailles decided to send serious reinforcements; at the time I was under the orders of the Baron of Vioménil, who led the legion of Lorraine. Antoine-Charles de Vioménil was born at the chateau of Fauconcourt, two leagues[7] from the château de Villé. Our properties were practically adjoining. Twelve years older than myself, he had proven himself during the Seven Years' War; in addition, his family had been presented at court, which led to a remarkable military career. I had the pleasure of joining him on all of his campaigns.

The landing was made at San-Fiorenzo[8] at the end of March. On the 1st of May, Monsieur de Vaux, who was leading the expedition, decided that we would camp at Oletta, a village located two leagues from San Fiorenzo. There were

---

[7] Approximately 10 kilometers or 6.2 miles.

[8] Saint-Florent today.

twenty-three battalions there. The other half of the army, made up of corps that had arrived the preceding year and were under the orders of Monsieur de Serent, settled on the plain of Mariana, not far from Burgo.[9]

The two troops advanced quickly, to the great surprise of the Corsicans, who did not expect us to be able to move so fast with our train of artillery on such difficult terrain. On May 14th, our legion from Lorraine joined the troops of M. de Serent on the banks of the Golo River. We knew that the Corsicans were entrenched upstream, near Ponte-Novo,[10] to block access to Corte, the Corsican capital. Four defenses protected the bridge. On the 16th at 5:00 in the morning, we left camp, crossed the river at a point we had found the night before, and went up through the woods to take the defenses from behind. Several of our battalions appeared near the bridge and two field pieces fired at them. The enemy, attacked from the front by the majority of our troops, and seeing that our legion was also arriving from behind, withdrew their defenses from above. The whole division could therefore cross the bridge quite easily.

We lost only about 30 men, killed or wounded, while this bridge had cost the lives of more than 5,000 Germans when they had tried to take the island. The actions of Baron de Vioménil and our legion from Lorraine were essential to this success.[11] What would be known as "the battle of Ponte Novo" marked the end of the Corsican conquest. The combat continued, but the leader of the Corsicans, Paoli, saw that he could not resist the French Army, so chose the path of exile and left for England.

---

[9] Borgo today.

[10] Ponte Novu today.

[11] Lallement, L. (1889) *L'expédition de 1769 en Corse et son annexion à la France.* Bulletin de la Société Polymathique du Morbihan.

THE POLISH CAMPAIGN

On December 30th, 1769, I was appointed captain. I was 29 years old. A year and a half later, in August 1771, I left for Poland with the Baron de Vioménil.

For some years, the Duke of Choiseul, then the main state minister for King Louis XV, tried to limit the ambitions of the Russian Empress Catherine II concerning Poland. On their side, in Bar, the Poles formed a confederation that brought together those who wanted to fight against a Russian invasion.

To force the Russians to split their army, and therefore help the confederates, the Duke of Choiseul pushed the Ottoman Empire to take up arms against them. In 1768, he sent Mr. de Taulès to Poland with a large sum of money to put at the disposal of the Poles to support them in their combat. He returned to France with the funds, deciding that there was nothing to expect from the confederates who were, in his view, just a group of gentlemen without order or discipline, commanded by leaders who were always divided.[12]

In July 1770, the Duke of Choiseul then commissioned Dumouriez, and bestowed upon him not only funds, but approximately thirty officers, including two for artillery, two engineers, and six gunners.[13]

Dumouriez failed in his mission and quickly irritated the Polish nobles. When the Duke de Choiseul fell in disgrace, the duke d'Aiguillon decided to replace Dumouriez with the Baron of Vioménil to try to assist the Poles in their fight against Russia. I was pleased to accompany him to Poland where we arrived in early September 1771.

Our first concern was to understand the military situation and the forces the confederates had. First a considerable array of Polish forces, well trained and disciplined, were shown to us. Then Dumouriez described for us the deplorable state in which he found the confederation, the way he had temporarily corrected the situation, and finally, the desperate state it had been reduced to through its misconduct. This way of explaining the facts was typical of Dumouriez, who often attributed successes to himself and failures to others. Nevertheless, his bleak picture of the Polish forces was not far from reality.

The troops were impoverished and undisciplined, without consistency or order. The soldiers had not been paid, were practically naked, malnourished, poorly armed, and even more poorly drilled. The confederation only had a few

---

[12] Letters of the Baron de Vioménil.

[13] Memoires of Charles-François Dumouriez.

places inadequately fortified and ill-supplied.[14] We first met with the main leaders to reestablish a minimum of order and discipline, expanding the garrisons by increasing recruits, paying officers and soldiers, providing supplies, and repairing and expanding fortifications. We had with us only about twenty officers, but they were all of great value, particularly Monsieur de Choisi, who we had send to Tyniec, a village near Krakow, to see if it would be possible to drive out the Russians.

In February 1772, Monsieur de Choisi managed to seize the chateau of Krakow with six hundred men. This was greatly lauded by everyone and boosted the morale of the Poles. The city of Krakow remained in the hands of the Russians, who immediately organized a blockade of the château. Mr. de Choisi resisted the repeated attacks by the Russians until April, when he finally had to surrender. He was surrounded by 10,000 Russians and we had not found a way to bring him aid.

During this time, we remained close to the main commander, who was supposed to organize and coordinate the confederates. Unfortunately, some of the leaders did as the pleased and the Polish forces missed several occasions for decisive victory over the Russians. The Poles paid very dearly for their disorganization and lack of discipline. On July 25th, 1772, Russia, Austria, and Prussia signed a treaty to divide up Poland, with Russia claiming several provinces.

The confederates lay down their arms in August and the Polish prisoners were sent to Siberia with their families, with others to follow. Those who could went into exile, mostly in France.

My military career continued. On the 24th of March, 1772, at the age of 32, I became a lieutenant colonel and left the engineering corps. A letter from Monsieur de Monteynard dated October 10th informed me that, based on the testimonial of baron de Vioménil, I was a made a knight of the Royal and Military Order of Saint Louis.[15] I was very pleased by this distinction, but I did not know that it would be eight years before I would again find active employment in the army.

---

[14] Letters of the Baron de Vioménil.

[15] Letter of October 10th, 1772, signed Monteynard and Vioménil.

THE WAR IN AMERICA

While I was completing my military studies at La Fère and Mézières, great powers were clashing in what was known as "The Seven Years' War." In the east, the kingdom of Prussia opposed Austria for the possession of Silesia. France, an ally of Austria, helped in this war, which was extremely costly. We were also at war with England, who wanted to limit our colonial ambitions. The immense territory of Louisiana, which stretched from the region of the Great Lakes to the Gulf of Mexico, was at the heart of the conflict between France and England, each country wishing to appropriate the revenue from the fishing and fur trades. The English also wanted to combat the influence of the Catholic church in that part of the world.

Compounding the error of doing battle on two fronts was the incompetence of the generals. Their presence was due more to the intrigues led by Madame de Pompadour than their military value. The French court, which Louis XIV had originally brought together in Versailles to rein in an earlier rebellion by the dissatisfied high nobility, was one of the main reasons for the ruin of the country: laziness, distance from the true reality of the country, the isolation at Versailles, lavish spending, exorbitant pensions paid out to favorites, the intrigues and the power of the mistresses of Louis XIV, and Louis XV squandering the immense riches of our country.

It was therefore England who emerged victorious from the Seven Years' War, making enormous gains as they did so. Even so, the many compensations would not be enough to refill the coffers of the British Crown emptied by the war, so the English imposed new taxes on the American colonists. In exchange, the colonists asked to be represented in the House of Commons. London refused, and relations began to get tense. At the time, the American colonies had the same political institutions as England; the communities decided their local affairs, and formed the counties that made up the state. The counties named representatives to the general assemblies of the states. The governors were appointed by London, but it was freely and at the request of Whitehall that they voted the subsidies to be given to the motherland.

In 1765, the British Parliament decided without consulting the colonies to highly tax the paper used for drafting acts, violating a right that had been hitherto respected. A congress composed of deputies from all of the colonies took place in New York and resolved, while remaining loyal to the Crown, to defend their liberties to the end. The Americans therefore broke their trade relations with England. London gave in, while reminding them that the legislative power of

Parliament extended to all parts of British territory. It was by this principle, during the summer of 1769, that new taxes were introduced. The Americans, calling on the principle that no citizen is required to pay taxes which were not voted by their representatives, refused to pay them.

In addition to the taxes, the thirteen American colonies were forced to sell their goods exclusively to England, and certain merchandise had to be purchased from England. These obligations were very expensive for the Americans. The tea tax was one of the most exorbitant and it became a symbolic point of disagreement between England and its colonies. John Hancock, a privateer captain, organized the importation of smuggled tea and a boycott of tea sold by the British East India Company. The company's sales in the American colonies plunged. The British Parliament then voted the Tea Act, authorizing the East India Company to sell its surplus tea to the colonies without paying tax. This measure, which allowed the company to sell its tea cheaper than the other importers and smugglers, ruined the independent merchants. Philadelphia and New York refused to accept the cases of tea sent to them by the company, and on December 16th, 1773, sixty Bostonians known as the Sons of Liberty disguised themselves as Native Americans, climbed aboard the three tea-trading ships anchored in Boston, and threw their cargo of tea overboard.

In England, opinion was divided as to how to respond to this. Some, including Sir Edmund Burke, considered the act showed the desperation of the colonists, who had to pay heavy taxes when they were not even represented in Parliament. Others felt it necessary to punish this action, and they would have the last word. General Gage came and settled in Boston, the port was closed, and the local courts were dissolved. England also raised a large army to ensure the colony obeyed.

The colonists were outraged by this treatment in Boston. While volunteers ran to oppose the occupation by General Gage, the delegates from the colonies responded to the Intolerable Acts by gathering in Philadelphia on September 5th, 1774, to establish the First Continental Congress. In October, they adopted the Declaration and Resolve of the First Continental Congress, which included a bill of rights, and petitioned King George III. Like earlier petitions, the one had no effect. The American War of Independence began in earnest in 1775 with the Battle of Bunker Hill in Boston, after which King George declared parts of the colonies to be in an "open and avowed rebellion."[16] Then revolutionaries were forced to abandon their position but suffered fewer losses than the English (100 for the revolutionaries and 1,000 for the English, with a majority of officers) and the battle allowed them to show their courage and determination. "A few victories like this will surely end British domination in America," said the English General Henry Clinton of the battle.

General Washington was named commander-in-chief of the American Continental Army. At the time he had few men, about 5,000, who were poorly equipped, poorly armed, and not very disciplined. However, that did not stop him from winning important victories against 50,000 soldiers of the British army. The most striking was at Saratoga, during which he captured 10,000 prisoners, and which also provided him with a large number of rifles and cannons. The loss by the American army in Montreal, Canada was partly offset by the taking of Boston in March 1776 and the failure of the English fleet at Charleston in June the same year.

On July 4th, 1776, the thirteen American colonies declared their independence and laid the foundation for a new government where they were each represented, marking a definitive break with England. In April 1777, a young Lafayette chartered a ship to travel to America. He was introduced to General Washington, and this was the beginning of a father–son relationship between the childless 45-year-old general and Lafayette, who was 20 years old. It was at a moment when General Washington was in the most profound despair that, in his own words, "Lafayette came to dispel my dark thoughts as the dawn dispels the night." Not only did Lafayette dedicate part of his fortune to supporting the Americans, but he refused the compensation offered to him by Congress. He had great physical and moral qualities, a noble character, courage, frankness, and skills in combat that impressed General Washington and all of the patriots.

France was then ill-disposed toward the Americans and General Washington, who had fought with the English against them during the Seven Years' War. It was thanks to the intervention of Benjamin Franklin, then the American Ambassador in Paris, and to the successes of the insurgent army, that the king of France decided to come to the aid of the Americans. A treaty of alliance was signed on February 6th, 1778. The Americans were reluctant to have another country intervening on their soil, and the contribution of the French was at first

---

[16] "By the King, a Proclamation, For suppressing Rebellion and Sedition" https://www.masshist.org/database/viewer.php?item_id=818&img_step=1&pid=2&mode=transcript#page1

financial and maritime, aimed at hindering trade exchanges between England and its colony.

In addition to this official aid, the French fought alongside the Americans. They were volunteers who brought moral support, but they were not always appreciated, as many of them were adventurers, arrogant, and demanding high-level commanding positions, which was poorly viewed by the Americans. The young Lafayette stood out from them with his modesty, courage, and enthusiasm.

In 1779, the military, financial, and political situation of the "insurgents" deteriorated significantly. It was clearer and clearer that without direct intervention from France, the American army would not reach its objectives. The Americans therefore accepted the presence of French troops on their soil.

At this time, the abandonment of the French–Spanish plan to land in England freed up the French forces to be available for other engagements. The King charged the Prince de Montbarrey, Minister of War, to prepare the expedition and appoint its leader.

In early March 1780, I learned that the Comte de Rochambeau had received confirmation of his appointment at the head of the expedition. The young Lafayette had done everything to obtain this role, using all of his connections at court. His personality worked in his favor, but the minister Montbarrey preferred a count whose experience and wisdom were well-known. Time would show that this was the right choice.

The king instructed the Comte de Rochambeau to move under the orders of General Washington, which was understandable as our army was fighting on American territory. However, the situation would prove complex given the American army's lack of means and its relative inexperience. The intelligence, sense of diplomacy, and natural authority of the count allowed them to avoid certain errors which doubtless would have been fatal to the expedition.

The general staff of the Comte de Rochambeau was made up of brothers the Baron and the Count de Vioménil. The baron was the second in command, and the Chevalier de Chastellux served as major general and Count Wittgenstein. All were valuable and experienced generals.

I learned that the Comte de Rochambeau had asked the minister for "blank" nomination orders to choose who he wished, and that I was on the list of six officers chosen.[17] I had the honor of being named adjutant general and first aide-de-camp of the Baron de Vioménil, the third campaign I would do with him. I

[17] Trintignan de J. *La France au secours de l'Amérique.* p. 46.

was even more flattered as I knew that at Versailles, the choosing of officers for this expedition had led to ferocious struggles for influence and evoked feverish agitation at court. The protégés of the Versailles court had of course been favored, and this blank order had been the only way for the Comte de Rochambeau to be surrounded by officers of his choosing and to free himself of the intrigues of court.

The role of the Chevalier de Chastellux, as major general, was considerable.[18] He designated, with the other majors of the army, those troops who had to stand guard, went on expeditions, composed the detachments, and escorted convoys. His functions were constantly exercised as he had to watch over everything that happened in the army. He could see the general at any hour and was lodged near his quarters because every evening he received instructions to pass on to each brigade major. For this he had an orderly sergeant per brigade to transmit orders. On the day of combat, he received a plan from the general of his army and the disposition of the regiments.[19]

As his adjutant general, I was his right hand, a great honor, but also a heavy task. I was also the first aide-de-camp of Baron de Vioménil, along with his nephew d'Olonne, whose family we knew, also from the Vosges.[20] The count informed us that we would receive 1,000 livres per month, a large sum to us, but that 200 livres would be withheld to pay cadets and draftsmen.[21]

---

[18] Head of the general staff of the chief general.

[19] Trintignan de J. *op. cit.* p. 217.

[20] The relationship between the d'Olonne and Ménonville families continues today.

[21] Trintignan de J. *op. cit.* p. 46.

THE CROSSING

When I finally arrived in Brest at the end of March, the fleet was complete and the belongings and provisions for the 5,000 men initially planned were being loaded on board. The troops, their equipment, artillery both for field and siege, and everything the army needed was loaded onto about thirty vessels, escorted by ten warships.

On the docks, the provisions for the voyage were gathered: sheep, poultry, salted beef and pork, and also barrels filled with biscuits, codfish, cheese, and rice, not to mention water and wine.

There was not enough space to load everything. One part of the contingent would have to wait, under the command of Count Wittgenstein—a second convoy that would never arrive. The initial plan was to bring horses, but they took up so much room that the Comte de Rochambeau decided to send all of his horses back; this served as an example, and he forbade the loading of anyone's personal horses. The wisdom of this decision was made clear when Admiral de Ternay, in charge of the squadron, explained that he had brought horses on the *Belle Poule* and, doubtless due to a long period of forced immobility, most of them were dead after only thirty-six days, and he could not use the surviving horses until six months after they arrived. It was finally decided to not take any horses, and to acquire what we would need once we arrived.

Finally, and despite the decision to leave a part of the contingent on land, the total number of passengers was not 5,000, but almost 6,500. It seems that among the 5,000 passengers initially planned, they had not counted the domestics and aides-de-camp who accompanied the officers. Each officer had at least one domestic, some like the Chevalier de Chastellux had up to eight, and the Comte de Rochambeau had twelve, including four who accompanied him on the *Duc de Bourgogne*.[22] The situation would create significant problems for everyone. Each vessel was given more people and supplies than it could reasonably hold, and some were quite worried about the conditions they would find on board. They were right to be worried.

On April 14th, I boarded the *Conquérant*.

Lodgings had been prepared for the officers. It was impossible, given the lack of space, to have a cabin for each person, and ten of us slept in a space where bunks had been set up. Some who were not officers, like Blanchard,[23] had to be

---

[22] de Trintignan J. op. cit. p. 19.

[23] War Commissioner, author of the *Journal de campagne* cited below.

content with a space made up between two cannons and a bed made from canvas stretched over a frame. Like all warships, the *Conquérant* only had two cabins, which were reserved for the commander and his second. My situation was no less comfortable than in the many camps I had known during the campaigns in Corsica and Poland. Fortunately, like every officer, I was able to bring a domestic with me, which made life as pleasant as possible. I was with the seven officers of the Saintonge regiment, the Baron de Vioménil, his two aides-de-camp, and his secretary. There was also Blanchard, the assistant to the War Commissioner of Tarlé.

I have to say that the comfort on these ships was rudimentary to say the least; it was almost impossible to be alone. We were fourteen to take our meals together with the commander,[24] in a small, low-ceilinged dining room, in a constant hubbub and among the smells of men and the dogs, which were essential to hunt the rats. Despite the dogs, the provisions were ruined quickly by the rats and the water acquired a dreadful taste.

Built in 1746 in oak with a pine rigging, the *Conquérant* was a vessel of seventy-four cannons and measured 170 feet.[25] On board, counting the crew, the passengers, and the domestics, there were 960 people and their supplies for six months, although the ship was built for a crew of 740 men.

Also on board were a surgeon and a chaplain, who were part of the ship's staff. The presence of the chaplain was required because life was regulated by prayers and services: morning prayer at sunrise, mass on Sundays and holy days. The chaplain also had to teach catechism, visit the ill and dying, and, in the case of battle, give his blessing and encouragement to those who were going to fight for God and the king.

On the *Duc de Bourgogne*, only the Comte de Rochambeau had his own room. No one else was entitled to this—not colonels or even generals. The situation was very distressing to those I call the "court roosters," who were very concerned with their personal comfort, and some offered him twenty-five louis to be able to occupy his room. The fate of the soldiers was much less enviable. They slept two by two, head-to-toe in hammocks, and since there were not enough of these, there were always half of the men standing guard. Instructions were given that the men stay the least amount of time possible in steerage, where the air was hardly fresh, and that they should take the air on deck.

[24] de Trintignan J. *op. cit.* p. 104.
[25] About 55 meters.

Model of the Conquérant. *Museum of Aquitaine, Bordeaux.*

Until May 2nd, the winds blew from the west, preventing us from leaving the bay of Brest. We only really took to sea on May 3rd, and soon the Bay of Biscay bestowed us with a storm that last four days and moored us for the rest of the trip. The English had initially planned to block us in the bay of Brest and, if possible, attack us there. Fortunately, the English squadron charged with this mission had to stop in England to get back in condition to sail and would only arrive on June 8th, long after our departure.

Our destination remained secret. We were headed toward Jamaica or Santo Domingo, and it was only on May 28th that General de Rochambeau announced that we were sailing to New England. This route by the south was more secure, as there was less risk of meeting an enemy squadron and we also had more favorable winds during our crossing and when we arrived near the American coast, where the prevailing winds came from the south.

The crossing was without major incident, if it was not for the difficulty of keeping forty or so vessels of all types moving. The most painful part for me was the forced idleness. Conversations with the other officers and the spectacle of the sea and the squadron in motion were the main distractions. Every day, we had to heave to and wait for the convoy that sailed less well than the warships. This tossed us about and made us seasick.

The only event worth reporting is the combat that took place on June 20th, west of Bermuda. On that day we met an English fleet, and Admiral de Ternay decided to regroup the convoy and put it under the protection of the frigates. Two ships, the *Neptune* and the *Éveillé*, received the order to go meet the enemy, which was made up of five warships and a frigate; for our part we had seven warships and three frigates. The English, doubtless baited by the size of the convoy, headed straight toward us. Monsieur de Ternay called back the two ships sent out as scouts, considering it more useful to defend the convoy than pursue an isolated English ship that was probably intended to distract and separate our forces. The two battle lines were formed. The two squadrons passed each other at long range and exchanged cannon fire. The English did not come any closer, and Monsieur de Ternay decided not to go after them.

The young officers immediately criticized their leader for not taking advantage of the situation to destroy this enemy force that was clearly inferior. I discovered that this "carrion" treatment of the commander or admiral, consisting of criticizing one's superiors, was very common in the officer's quarters. These reproaches were taken up and expanded by the "court roosters," who were always eager to engage in battle. I defended the admiral, considering his mission was to bring his fleet to port undamaged. The loss of several vessels, always possible in such an engagement, could have had grave consequences. This decision was certainly taken in agreement with the Comte de Rochambeau, who was concerned with the success of his mission above all. He and the admiral were bound to respect the instructions given to them by the king, with the least human and material loss possible. Moreover, the general later confided in us that he thought it would be impossible to conduct that battle with greater attention and skill.[26]

We did not know it then, but we could have paid very dearly the time spent on these battles, because we later learned that a large English fleet was looking for us. It reached New York the day after we arrived in Newport, and Rhode Island only a few days after us. Instead of peacefully landing in Newport, we would have been doing battle with a squadron much more powerful than ours.

Aside from these minor incidents, which gave us the opportunity to discuss amongst ourselves, we were not very busy and the conversations were repetitive, mostly about the distances covered—thirty to forty leagues a day[27]—and the number of sick and dead. Scurvy hit our ship hard.

The *Conquérant* had set sail on February 8th, long before the others, and the crew had been at sea since then, which explained the high number of sick people. Since June, we had already thrown thirteen bodies overboard and sixty more were ill, most of them with scurvy. Some of them had to be taken onboard the *Fantasque*, our hospital ship. In total, the *Conquérant* lost about one hundred men, but it is difficult to know the exact number of dead, as they were discreetly thrown overboard through a porthole.

At the beginning of July, time seemed to move slowly. The number of patients increased. Our desire to reach land was increased by the frequent fog, not knowing where we were, and the possibility of an English squadron with a mission to stop us from reaching Newport.

On July 9th, after sixty-nine days at sea, I finally clearly saw land ahead. We were three leagues from the small island of Martha's Vineyard, located twelve leagues[28] from Rhode Island. The general gave the order to drop anchor and set sail the next day. A resident of the island came and proposed to lead us there; he informed us that the Americans were still in charge of Rhode Island and that, as far as he knew, the English did not have large naval forces in the area. He told us

---

[26] Comte de Rochambeau. *Mémoires militaires, historiques et politiques.*

[27] About 120 to 160 kilometers, or 75 to 100 miles.

[28] About 46 kilometers, or 30 miles.

rather naively that he was not a royalist or a revolutionary, and that he wanted to be friends with everyone.

NEWPORT

On July 1st, around 7:00 in the morning, the fleet dropped anchor a quarter of a league from Newport; the general disembarked the same day. Our first contact with America was quite a surprise: the small town of Newport was deserted, the inhabitants closed up in their homes. This was a great disappointment to us, as we expected to be welcomed as their liberators.

Later we were given two reasons for this welcome. The first was the negative experience of when the English troops had occupied the town; the inhabitants had no desire to relive anything like it. The English occupation had been particularly odious: many houses had been destroyed, churches has been rendered unusable, neighboring farms had no more livestock, and the inhabitants feared the same problems with the French. The second reason could be credited to Admiral d'Estaing. In August 1778, an army of 9,000 Americans, commanded by General Sullivan, and 4,000 French from Admiral d'Estaing's squadron was well positioned for success in an attempt to reconquer Newport. The surrender of the English was close, when on the horizon the Admiral Howe's fleet appeared. D'Estaing, eager to confront the enemy to show his courage and receive the honors of a victory, raised anchor and went to meet the English. A storm, which would last for two days, forced them to suspend the battle to repair the damage. D'Estaing refused to return to Newport, despite the insistent requests of Sullivan and Lafayette, and decided to go to Boston, leaving Howe to protect the town once again. The assailants had to retreat, a retreat that became a rout. D'Estaing was greeted in Boston with insults, and there was talk of treason. We therefore understood why our arrival did not arouse great enthusiasm.

I have to say that pride and the desire to shine and be the object of admiration at court were powerful motives for our officers. Too often, this caused them to lose their reason and led to many military failures. The behavior of Admiral d'Estaing that day was just one example.

The day of the 12th was dedicated to finding a campsite for the troops. It was urgent to get them off the ship so they could get exercise and change their diets. Once a place was found, the general successfully reassured the population about our intentions and the behavior of our troops. He made it clear that rent would be paid for any occupied spaces, not in paper money but in pieces of gold and silver. This promise started to reassure the inhabitants. We then started to look for provisions, always saying that the French would pay for anything we took. The magic of the money worked, and the shutters and doors began to open. That same day, early in the morning, an envoy of General Washington arrived to explain to the inhabitants that the French were indeed there at his request, and not for personal gain, as the English had told them. Rochambeau's letter to Washington, placing himself under Washington's command, also had a positive effect. The personality of the count—warm, simple and hearty—finally reassured the residents. In the evening, the general opinion had completely turned around, and the town was brightly lit, making a beautiful and welcoming tableau for those still on board the ships.

Vernon House in Newport, Headquarters of Comte de Rochambeau.

It proved fairly simple to find places to stay. I was in Black Street[29] at the home of Captain George. His house was adjacent to the one where the Chevalier de Chastellux stayed, because I had to be able to go to him at any moment. Baron de Vioménil and his brother were in a street just behind[30] and the Comte de

[29] Spring Street n° 90 today.
[30] Thames Street today.

Rochambeau stayed in New Lane[31] where he had acquired Vernon House[32] to use as headquarters. All of the staff was therefore grouped around the general, as these houses were close to each other.

Our first task was to organize the camps, the different regiments, and especially our defense. Lauzun's legion immediately took a position facing the port, to be able to see any ship approaching the coast and to block enemy ships access to the port. From to July 13th–15th the troops came off the ships and camped in an orchard.[33] The Americans who came to visit us were surprised by the behavior and discipline of the soldiers who would not even take the apples hanging over their tents without having paid for them. The news that the French were a blessing for the inhabitants quickly spread.

Throughout our stay in Newport, maintaining order and discipline was a constant consideration. Among the officers, there were so many duels they had to be sanctioned. They were mostly held over gambling debts, questions of military honor, but sometimes the reasons were trivial; once two men were fighting because they could not agree on the amount of time to cook a chicken![34] Among the troops, it was theft that needed to be punished. Thus, a servant who took a piece of cheese from a farm without paying was sentenced to twenty-five strokes with a rod,[35] and two others who had stolen silver cutlery[36] were executed. These sanctions, however, were exceptional.

The priority was to dig trenches and fortify the point where an enemy could attack. Nevertheless, the army needed two days of rest before starting to work. While the artillery was being unloaded, we saw about twenty sails headed toward us. It was easy to see they were enemy sails, which caused a great commotion, as we were not yet ready to defend ourselves effectively. This squadron belonged to General Clinton, who had embarked from a port on Long Island with 10,000 of his best troops, a lot of large artillery, and mortars for attacking us. He spent a few days off the coast of the island. Then, in learning that General Washington was approaching New York, and fearing that the garrison left there would be insufficient, the English decided to pull back. An attack by the English at that time would certainly have caused us considerable damage, given the state of our

---

[31] Clarke Street today.

[32] *Magazine of Newport History,* p. 90.

[33] Bellevue Avenue today.

[34] Bodinier, G. *Les officiers de l'armée royale.*

[35] Diary of Comte de Rochambeau. Transcription by Arnaud Blondet.

[36] Bodinier, G. *Dictionnaire des officiers généraux.*

defense. The English fleet reappeared several times over the summer, but we had consolidated our defenses and were ready to meet them.

At the request of General Washington, Boston and Rhode Island had sent us 4,000 to 5,000 men as reinforcements, under the command of General Heats. We only kept 2,000 of them, placed under the order of Lafayette; the count decided to send the rest back to their harvesting, which they had been generous enough to interrupt to come to our aid.[37]

It was also urgent to organize hospitals to treat our patients. When we landed, we numbered 600 to 800 the land troops, and 1,000 to 1,500 navy troops. In a letter dated October 1st to the Prince de Montbarrey, the Marquis de Chastellux announced that 186 soldiers had died since May 1st, including 153 since we had landed, most of them from scurvy and dysentery caused by the summer heat.

After the forced inactivity of the crossing, this was a period of intense activity for me. The practical problems of setting ourselves up were at times difficult to resolve, as we were lacking almost everything and there was no store, market, or garden. The difficulties were even worse for those who did not speak English, which was the majority. I silently thanked my instructors; it was thanks to them that I spoke fluent English.

The English who had left the island ten months earlier had destroyed many houses and trees before they left, giving the town an air of desolation. If the residents were to be believed, before the war the island had been a delicious place with an agreeable climate and fertile soil. Few fields were cultivated when we arrived, as many owners had abandoned them.

The port of Newport, located at the east end of Rhode Island, was superb and large enough to welcome several warships. In the south it was defended by Fort Brenton; located on an outcrop, this fort had been refurbished and equipped with twelve 24-pounder guns.[38] On the other side of the channel, the small Goat Island had forty 36-pounder long guns. This situation, complete with a second island—Rose Island—and the warships, allowed us to take on any ship trying to enter the port in a powerful crossfire. A battery of eight 8-pounder guns had also been placed on the northwest of the island, facing the mainland at its narrowest point, to block the passage of a fleet while protecting the arrival of reinforcements if needed.

---

[37] Comte de Rochambeau, *op. cit.*

[38] Firing cannonballs weighing 24 pounds, or about 12 kilograms.

As for the city of Newport, it was located along the port and had two main avenues. The houses were built of wood and were small but charming, painted in different pretty colors. Each owner decorated the interior to his taste, and often there was not much furniture but everything was perfectly clean. Some houses were built outside the town and placed onto wheels to be brought to their final location.

Map of the city and harbor of Newport.

The lifestyle of the inhabitants was fairly uniform. Tea, the favorite drink, was very strong and barely weakened with a dash of milk. It was generally served by the mistress of the house between 4:00 and 5:00 in the afternoon, and she offered it to everyone present. Her insistence was such that it was difficult to refuse. The coffee was very light, almost the same color as the tea, and also served with a dash of milk. Breakfast was composed of tea or coffee, toasted bread and butter with ham, cheese, and sometimes fried bacon. Using sugar marked the difference between poverty and abundance. Dinner was made up of boiled meat with boiled vegetables, served in dishes in impressive quantities and all covered with a brown gravy. A bowl of cider, beer, or grog made from rum was available for those who were thirsty. There were no glasses, only the bowl that was presented to everyone, and all drank from it. When I arrived at someone's home, the master of the house always offered me a drink. He drank to my health first and then it was my turn. After dinner, in the most prosperous homes, the table was cleared, and the ladies left the men to continue to talk. Madeira wine was served, and the men spent a long while drinking, smoking, and talking. In the evening, a light supper was taken around 10:00.

During our dinners between Frenchmen, once we had covered the military subjects, we share our impressions of the Americans and especially American women. The general view was that the men were rather tall and well built, but did not give an impression of great vigor, doubtless due to their pallor. The women were also pale and seemed fragile. They aged prematurely and at 20 years old looked like they were 30. Their complexions were very beautiful and very pale, their faces pretty and sometimes beautiful with delicate and regular lines. They were even more beautiful when dancing had given them some color.

Over time, a certain trust was established. I discovered that the English had not painted a pretty picture of us: they described us as the worst people on Earth, old and ugly dwarves who fed on frogs and snails. However, the image eventually faded away and relations between us improved. But trade remained difficult and they offered products to us at exorbitant prices. It was impossible for us to find a house to live in or land to camp on without them asking for money. Horses were sold to us at double the usual price and in such a poor state that we had to refuse half of them. Not only had we to pay a high price to meet our needs, but we also had to maintain the American militia who arrived in Boston without tents, poorly armed, and without ammunition or provisions. Between ourselves, the Comte de Rochambeau said he regretted to see this behavior on the part of the people we had come to help, but he never complained about it publicly or made the reproach to our friends. I thought sadly that during this time, France was becoming so poor that the most disadvantaged were going hungry. Indeed, a few years later we would pay dearly for our generosity: the price of a revolution.

At the time, the inhabitants of Newport were divided into Whigs and Tories. The first were the good Americans, who fought to liberate their country from the English yoke, and the others were the loyalists or royalists, loyal to the king of England, and made up three-fourths of the inhabitants. Even though the Tories had supported the English, who had engaged in a great deal of pillage and violence, all were nevertheless invited, without discrimination, to our balls and those organized by Madame Cowley. Loyalty to a king did not seem to be a major fault to us; we even felt a certain sympathy toward the loyalists. It was the wives of the Whigs who were the most intransigent, refusing to dance in the same room with the wives of the Tories, or even to attend our parties if the Tories were invited. As there were more of them, we invited them anyway, and thus had no shortage of escorts and our receptions were always a success.

There were different religions living together in Newport, which therefore had several churches. There were many Protestants but the main religion was Anglicanism. The Quakers were also present in the city and on the island. These men were grave, their manner and clothing austere. They spoke little, used formal address when speaking with everyone, and never took off their black hats, even when entering a house. Their religion was based on the fear of God and love of one's neighbor. They disapproved of all bloody acts and were therefore disinterested in war, to the point that they refused to join the parties organized to celebrate the success of their country. This could be explained by the fact that they were Tories. They condemned slavery and felt great solidarity among themselves and generosity toward others. They had no priests or officiants and were happy to meet twice every Sunday in a common room open to all. There, men and women were separated, and there was a great silence; each person was seated and could speak if they wished. It was easy to see who was going to speak, starting with convulsive movements and a voice and body that began to shake (or "quake," hence their name). Sometimes the gathering ended without anyone speaking a word. Their religion was austere and no pleasure was tolerated—no singing or dancing. It seems these prohibitions particularly displeased the young women, who were rather pretty and cheerful.

As far as the military situation was concerned, it was delicate, to say the least. General Washington, we later learned, was extremely embarrassed by the state of his army. Many soldiers went home to take care of their harvests; there were few men and little money. He could not make a serious contribution to a campaign and was therefore uninterested in meeting with the Comte de Rochambeau. He therefore dispatched Lafayette, who would arrive on July 28th in Newport to meet General de Rochambeau and Admiral de Ternay. On August 9th, when Lafayette had joined General Washington, the count received a letter from Lafayette urging him, in the name of General Washington, to undertake certain campaigns and in particular to attack New York even though we had no navy. The general answered, politely but firmly, and began by reminding him that there were 14,000 troops settled in the New York islands and that the French Navy was blocked in Newport by a fleet one and half times its size. The count said that as soon as the French troops left, it would be easy for Clinton to destroy our fleet and attack us.[39] The count also told him that only he could judge if his men were ready for battle, and that he did not wish to risk the lives of his men in a foolhardy manner. Moreover, failure would have disastrous consequences on the army's morale for the rest of the campaign. At headquarters, we were all the more unhappy with the tone of Lafayette's letter, as, in the same letter, General Washington wrote to General de Rochambeau and did not mention this plan at all. The count also wrote to General Washington asking him to give him his orders directly in the future.[40] General Washington recognized the strength of the count's arguments, who, from the moment their exchanges were direct, could only compliment the soundness of the General's judgment and the graciousness of his style. This would be the beginning of a long correspondence between the two generals.

It must be said that the young Lafayette had loudly proclaimed to the court that he was the ideal man to command the expedition, and the decision of the king and his Minister of War, Prince de Montbarrey, to name instead the Comte de Rochambeau had greatly disappointed him. Upon his return to America, he thought he could take over as head of operations by convincing General Washington to entrust to him his relations with General de Rochambeau. After this rebuff, Lafayette no longer tried to interfere and General Washington gave him command of the American troops in Virginia and the Carolinas. This mission was well suited to the young Lafayette, who, even if he was prone to grandstanding, was a wise and brave soldier in the field. He seemed to me to be more a man of action than politics, and his behavior during the Revolution would confirm this. The same letter of August 9th would bring a letter from General

[39] Comte de Rochambeau, *op. cit.* 66

[40] Letter from Rochambeau to Lafayette.

Washington, not mentioning at all an attack of New York but explaining that his presence at the head of the troops in front of this city was essential because an English offensive could take place at any moment. In addition, his presence in New York had the advantage of preventing the English from weakening their defense by taking away troops that would be needed for action against Rhode Island.

At the end of August, we received a visit of about twenty Iroquois people. One of them, who came from Canada and spoke French, explained to us that most of the tribes helped the English, providing them with alcohol and rifles.

As for the visitors, they wanted to be friends with the French. The general gave them several gifts, such as medals, swords, shirts, and red wool blankets. He also let them know that he would ensure that they did not run out of brandy. The next day, the general had his army put to arms to show the Iroquois their numbers and strength. He also showed firing maneuvers with powerful military salvos. It all seemed to impress them greatly but they were especially interested in our carpenters and their axes. In the evening, they demonstrated one of their war dances, and how they scalped their enemies for trophies. They were almost naked, and their skin, naturally a tone close to copper, was painted in different colors. Their noses were pierced and medallions hung from them; their earlobes were also pierced and some reached down to their shoulders. They coated themselves with a pungent-smelling oil.

The Americans suggested that it was not a coincidence that the Iroquois had come to visit. They usually fought with the English and doubtless wanted to see the size and strength of the French troops in order to choose to fight on the winning side. Was this true? In any case, after this visit the Iroquois did not aid the English and remained neutral.

At the beginning of September, we received bad news from the American army. Lord Cornwallis had been in the South ahead of General Gates; Gates was beaten and the American army was completely routed. De Kalb held off the English with one division, but was killed. General Gates withdrew with what was left of his army to Hillsborough, North Carolina.[41]

The Comte de Rochambeau finally obtained from General Washington the appointment he had wanted for so long.

---

[41] Balch, T. *Les Français en Amérique pendant la guerre de l'Indépendance des États-Unis: 1777–1783*. p. 105.

General George Washington in Princeton, by Charles Wilson Pearle. Pennsylvania Academy of Fine Arts.

Portrait of Comte de Rochambeau by Charles-Philippe Larivière. Château de Versailles.

The interview was set for September 20th in Hartford. General Washington proposed to attack New York, specifying that it should be done with the French navy alone. The count, who awaited at least two regiments that had been promised to him, explained that he did not have sufficient forces for such an attack. He also had the skill to establish the principle, which General Washington unreservedly accepted, that no action would be undertaken without the control of the seas being assured on the coast where the action would take place. By accepting this point, General Washington became even more dependent on the French, who were the only ones able to guarantee this essential control of the seas. Still, it was decided that as soon as the second French division arrived, or there was an increase in the naval forces sent by Mr. de Guichen, an attack on New York would take place. These hopes would be ruined when the generals received the news of the arrival in New York of Admiral Rodney's fleet, tripling the English forces.[42]

The Comte de Rochambeau returned to Newport, where he prepared his army to spend the winter. We had no news of the convoy that was bringing the troops who had stayed in Brest under the orders of Count Wittgenstein. It was too late to consider a campaign, so we prepared ourselves to spend the winter in Newport. The harshness of the climate made wintering under a tent impossible, so the Comte de Rochambeau proposed to the inhabitants that we repair, at our expense, the many houses that were in ruins and use them to house our soldiers. We would then return the houses to the Newport inhabitants when we left. They were very happy with this proposition, and for us it was a much less expensive solution than constructing new buildings. The soldiers went to work and were able to move in at the beginning of November, when the weather was becoming cold and rainy. The count also had them build, in the garden of his Vernon House headquarters, a house where the officers could meet and organize parties and balls where the residents could be invited. It was inaugurated in January.

It was at the end of September that the treachery of the American General Benedict Arnold was discovered. He had established secret communications with the English General Clinton, and planned to hand over West Point to him. This American stronghold dominated the Hudson River and its location was extremely important. Arnold had met Major André, an aide-de-camp of Clinton, to arrange the details of the operation. On his way back to New York, André was arrested by militiamen, who found the plans of the scheme in his shoes. As soon as he

---

[42] *Ibid.* p. 108.

heard that the plot had been discovered, Arnold fled and joined an English ship. This betrayal was all the more surprising given that General Arnold had always shown exemplary dedication and courage in the many battles in which he had participated; until this day he had been considered a hero of the fight for American independence. The saddest part is that Major André, an English soldier who had won us all over with his conduct, was shot.

During our stay, I had several occasions to go to Boston, which was about twenty-five leagues from Newport.[43] The route between Newport and Boston was completed in two or three parts. The first in general was Providence, which was one days' walk; it was a small town and quite pretty. When I was not on official business, I stayed there at inns that were always very clean.

The landscape resembled part of Normandy. We travelled through many forests and streams, and the farmland was surrounded by walls or wooden fences. There were not many people living in this region, and the few villages were only some scattered houses. It was a very pleasant journey.

The land was fertile because it had just been cleared. I thought it would be quickly depleted, but the colonists told me that when that happened, they would find other land further west. They were happy farmers who enjoyed generous nature and almost unlimited space. However, the English could not take it over because the lands were still inhabited by Native Americans for the time being,[44] and the English forbade the settlers to move further west.

Boston is situated on a peninsula connected to the land by a causeway.[45] It is defended by two strongholds and, on the two hills in the middle of the city, forts had been built to protect the harbor, which is superb.[46] One could see the lighthouse, which gave the alert in case of danger, and from there was a magnificent view: the city stretched out in front of you, as well as the port with its islands in the middle. A bit farther one could see the city of Charleston, which the English had burned down after their defeat. The city of Boston was formidable and must have been even bigger before the war. The houses were sometimes built in brick but more often in wood, painted as elsewhere in different colors and always comfortable and well maintained. The house of the governor,

---

43 Approximately 100 kilometers or 62 miles.

44 Blanchard, C. *Guerre d'Amérique. 1780–1783. Journal de campagne de Claude Blanchard.* p. 85.

45 This peninsula no longer exists, as the space separating the land has been filled in.

46 Balch, T. *op. cit.* p. 124 et 133 And Count de Clermont-Crèvecœur, J-F-L. *Journal de la guerre en Amérique pendant les années 1780, 1781, 1782, 1783.* p. 81.

His Excellency General Hancock, was very beautiful. Boston was above all a city of commerce, certainly one of the richest in the world, and had benefited from the arrival of many foreigners. Hides, lumber, flour, salted meats, and a great quantity of cod caught in Cape Cod Bay were traded here. The Bostonians were very busy with their business affairs, peaceful but inscrutable, and were among the coldest Americans I met, even if they became more gracious, or, if one prefers, more courteous with time.[47]

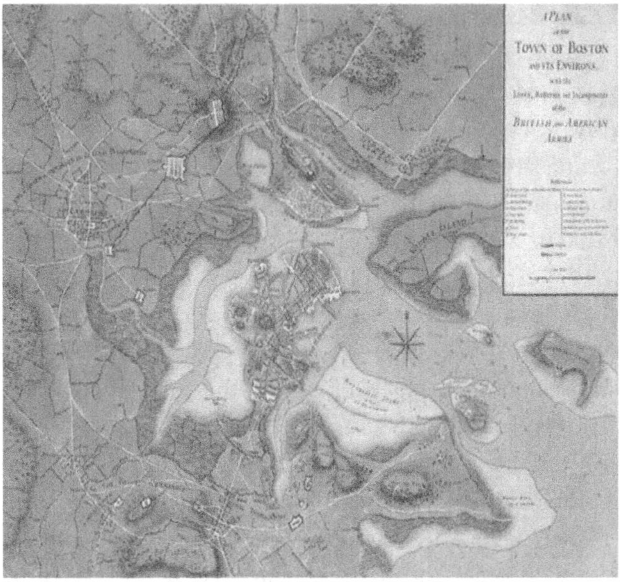

Map of Boston.

Seven miles from Boston, in the village of Cambridge, the Bostonians had created a university[48] composed of four brick buildings, a library of 5,000 books, and a beautiful printing press. The professors enjoyed an excellent reputation, were very concerned with the quality of their teaching, and corresponded with many European scholars.[49] The Americans gave great importance to their children's education, considering it an essential asset for themselves and their country.

[47] *Ibid.* p. 83.

[48] Harvard University.

[49] Abbé Robin, *Nouveau voyage dans l'Amérique septentrionale en l'année 1781 et campagne de l'armée de M. le Comte de Rochambeau.* p. 24.

On the military front, things were not going well for our allies. At the beginning of 1781, General Washington's troops were completely discouraged. The Pennsylvania Corps, based in Morristown, New Jersey, revolted on January 1st; the soldiers wanted to present their grievances to Congress, which were great and justified. Their general himself said they were poorly clothed, exposed to bitter cold with only threadbare coats and one blanket for three soldiers, poorly fed, and even more poorly paid.

Having obtained satisfaction, the mutineers returned to their camp, after having delivered to their officers the Englishmen sent to incite them to desert. A month later, hearing of the success of the Pennsylvania troops, the Jersey troops stationed in Pompton also revolted. It was necessary to severely curtail this revolt to prevent others following their example.

General Washington was convinced that a military success was needed to give confidence back to his troops. A joint expedition of the French army and navy was therefore sent to the Virginia coast. However, the combination of poor weather, bad luck, and poor coordination meant that nothing concrete was realized.

The expedition came back to Newport undefeated but also unsuccessful, despite the general, who sent a letter to one of his cousins at Mount Vernon where he sharply criticized the French admirals. The letter was intercepted by the English and given to Sir Henry Clinton, chief general of the British Army in America, who distributed it widely in New York and sent a copy to Comte de Rochambeau. The count was not personally attacked, but he could not allow these unjustified reproaches against his officers to pass. He therefore wrote to General Washington, first asking him if the letter was indeed from him and explaining why the reproaches were based on false information. General Washington did not deny that the letter had come from him, as he could have, but said that the letter was confidential and written to a relative. He recognized that his conclusions were based on false information and apologized. The incident was closed.

The Comte de Rochambeau wrote a note to Sir Clinton asking him in the future to write such letters to General Washington, the general in command, and not to him. Taking the high view, and the noble character of both men, had allowed them to avoid an incident that could have ruined the future of their relationship.

At the same time, a letter from the Chevalier de Chastellux to Monsieur de la Luzerne, French consul in Philadelphia, was also intercepted by the English and delivered to the count. In it, he criticized Monsieur de Rochambeau and boasted that he had succeeded in convincing him to attack New York; he complained about the count in rather dishonest terms. The count simply summoned the Chevalier, showed him the letter, and threw it into the fire. It was more proof of the nobility of character of the Comte de Rochambeau.[50] At least the letter convinced the English that we really did intend to attack New York.

A second meeting of the men took place in March 1781, when General Washington came to Newport. He went aboard the *Duc de Bourgogne* where the main officers were waiting for him. He then went back on land and reviewed the troops. He was visibly impressed and reassured by the order and discipline that he saw.

This visit was largely dedicated to the general meeting the main French officers; for my part, like most of my comrades, I found him to be very handsome. Tall, nobly made, well proportioned, his face pleasant, his countenance gentle and open, and his look attentive, noble, and assured. His manner was rather cold but always polite.

Later, I would have the opportunity, when my duties would lead me to meet and talk with him, to appreciate his even character, his seriousness, and his attention to whoever was speaking to him. Extremely modest, he accepted compliments graciously but seemed to avoid them rather than to seek them. Everyone who met him was impressed by this general who inspired respect without arrogance, in a very natural way.

He commanded his army perfectly, treating his officers politely but without familiarity. In return he received respect, trust, and admiration. He could seem slow to make decisions, but once he had made up his mind, he followed it with great courage. He was also capable of taking responsibility upon himself when the best interests of his country called for it. This is what he had done to settle the quarrel with General Gates. This man seemed to have all the best qualities, and we were all won over by him.

---

[50] Balch, T. *op. cit.* p. 129.

THE PORTSMOUTH EXPEDITION

The day after General Washington arrived in Newport, on March 6th, I boarded the *Duc de Bourgogne* with the Baron de Vioménil. It had been agreed between the two generals that while Lafayette would lay siege to Arnold in Portsmouth, a French fleet would attack by sea. To lead the expedition, the Comte de Rochambeau had designated Baron de Vioménil, accompanied by the Marquis de Laval and the Vicomte de Noailles, commanders of the grenadiers and light infantry; assistant quartermaster general de Collot; commissioner of supplies Blanchard; and myself. These choices were criticized by the main officers, in particular Lauzun and Laval, who saw themselves as best placed to lead this expedition.

Admiral Destouches was quartered on the *Duc de Bourgogne*. He brought seven ships of the line and three frigates—582 cannons total. We had 1,200 men on board, and artillery and mortars to establish a siege in case the expedition was successful.

As soon as we were out of the bay, a strong wind forced us to distance ourselves from the coast and disperse the fleet, which fortunately was rallied at the entrance of Chesapeake Bay. We then discovered the English squadron was composed of eight ships of the line and three frigates, totaling 660 cannons. The expedition was clearly compromised, but Admiral Destouches decided to fight, even if the enemy's forces were superior to ours. The first engagement was intense; we were with the Baron de Vioménil and Blanchard on the quarterdeck. Blanchard was visibly shaken by the combat; aware he was not a professional soldier, I offered him some tobacco and tried to make him laugh by saying that, contrary to what everyone said, this tobacco at least would not be bad for him.[51] I knew from experience that a bit of humor was an excellent remedy for fear.

The *Conquérant* had been seriously damaged, but the admiral decided to prepare for a second engagement. As for the English, concerned with protecting Portsmouth and having suffered more damage than we did, they entered the bay. This was a clever maneuver that allowed them to both take cover and prevent us from landing. We then returned to Newport.

For his part, Lafayette left with a detachment of 1,200 men. Having arrived at Annapolis, he continued the crossing with some officers in a boat and went through the middle of English frigates to reach Williamsburg and gather the militia there. He had already blocked Portsmouth and repelled enemy attacks

---

51 Blanchard, C. *op.cit.* p. 70.

when he heard that the English were blocking the bay. He could only return to Annapolis to rejoin his detachment. On this occasion, the young Lafayette once again proved his audacity, courage, and decisiveness.

MISSION TO GENERAL WASHINGTON AND CONGRESS

In addition to these military operations, we had to ensure provisions for the army, which was becoming difficult. In May, the Comte de Rochambeau charged me with a double mission. The first was to obtain the fulfillment of a contract concluded in France by Benjamin Franklin, the then American ambassador to France. This contract ensured provisions for the army and had been ratified by Congress, but it was difficult for him to respect it because food was scarce at the end of winter.

Bearing a letter of recommendation from the Chevalier de Chastellux[52] and another with instructions from the Comte de Rochambeau,[53] I met General Washington in New Windsor and our discussions lasted for several days. My mission was not an easy one, as I had orders to obtain satisfaction at any cost. We knew that General Washington did not have the authority to make a decision, but we needed his support with Congress. The general was difficult to deal with, but pleasant. He listened attentively and sincerely wanted the commitment to be respected. To be able to speak in English without an interpreter made it easier; our contact would have probably been even more cordial if I had been a freemason like himself, but I think the result would have been the same. Our discussions were followed up with letters[54] summarizing our points of view. In conclusion, the general said that the fulfillment of this contract was not his responsibility and recommended that I address my requests to Congress. For this he gave me a letter of recommendation for Mr. Samuel Huntington, the president of Congress.

My discussions with General Washington were rather harsh, but he knew that I was just respecting the instructions from Comte de Rochambeau. He must have remembered it in a positive light because years later he asked the Marquis de La Fayette and the Marquis de Saint-Simon to send me his "compliments."[55]

Equipped with a letter of recommendation from the general to the President of Congress, I left for Philadelphia, where Congress met at the time. After

---

[52] Founders Online, National Archives: letter *To General Washington from Chevalier de Chastellux,* April 29th, 1781.

[53] Founders Online, National Archives: letter *To General Washington from Comte de Rochambeau,* April 28th and 30th, 1781.

[54] Founders Online, National Archives: letters *From George Washington to François-Louis, Comte de Ménonville* and *From Comte de Ménonville to George Washington,* May 6th, 7th, and 9th, 1781.

[55] Lafayette, G du Motier. *Mémoires.* p. 152.

learning of the letter and our requests, the president introduced me to Robert Morris, the Superintendent of Finance. He told me that providing provisions for the French Army was not very promising, but he would do his best to satisfy our requests. Congress recognized the validity of the contract signed by Benjamin Franklin, but was simply unable to honor it. I therefore returned to Newport without any official guarantee from Congress. My efforts did have an effect, however, because, at the end of July, Robert Morris informed Benjamin Franklin that they would respect the commitments made to France.

Expedition from Portsmouth, by V. Zveg of the US Navy. US Navy Naval History and Heritage. François-Louis was on the Duc de Bourgogne *with Admiral Destouches and Baron de Vioménil.*

WE DECIDE NOT TO ATTACK NEW YORK

In May 1781, a third meeting between the generals Washington and Rochambeau took place in Wethersfield, Connecticut. The situation had changed considerably since their first meeting in Hartford. We now knew that we would not have more troops, not even the two regiments we had been promised, but we should have more money. In addition, Admiral de Grasse had left with a large fleet to the West Indies and the authorization to reach the North American coasts, if it did not endanger French holdings in the West Indies, and if he considered that their presence could benefit both the French and the Americans. As for General Washington, he had a small but real army, and we could consider a joint action.

The two generals agreed that it was essential to do something to raise the morale of the Americans, who were still very discouraged, and the French, who were beginning to wonder why they had come so far if not to fight. General Washington asked the opinion of the count, who suggested going south, driving out General Cornwallis's army, and destroying it. General Washington wanted to attack New York, which Comte de Rochambeau finally had to agree to. He was, above all, a soldier who respected hierarchy and the instructions that the king had given him. After this discussion, General Washington sent out several letters announcing his decision to attack New York. We learned later that, by chance, one of these letters had been intercepted by the English and given to General Clinton, who then decided to reinforce his defenses in New York. This interception was lucky for us, because the English troops were much more numerous than ours, and it is not sure that we would have won if General Clinton, instead of protecting New York, had decided to use all of his troops to attack us.

Our expedition to the south required the support of our fleet; General de Rochambeau therefore returned to Newport to convince the Count de Barras, who commanded the small fleet based in Newport, to join Admiral de Grasse's fleet and prepare the army's departure. To support our action along the coasts, General de Rochambeau had asked Admiral de Grasse to join us, preferably in New York, to give satisfaction to General Washington, who wanted to attack the city. The Admiral's response arrived on August 15th; he refused to attack New York because of a sandbank, which blocked entry to the port. The French ships, with a deeper hull than English ships, would need the assistance of port pilots to enter, but they were unlikely to provide support. In addition, New York was too far from the West Indies fleet, which was in charge of protecting it. However, they would be present at the entrance of Chesapeake Bay.

Finally, it was the count's plan that was accepted: attack Cornwallis. He had understood that the attack on New York would be suicidal, and had the tact to not push General Washington and let him come around to his point of view. I never knew whether Admiral de Grasse's decision not to go to New York was spontaneous or if it had been taken with Comte de Rochambeau.

THE CAMPAIGN BEGINS

On June 16th, Baron de Vioménil conducted an entry review for the campaign, and two days later, the troops left Newport. My duties as adjutant general increased even more. With the Chevalier de Chastellux, we had to regularly report to the count on the situation of the troops and their progress. Also, and especially, we had to plan the marching orders for the next day, refer them to the count when useful, and through the aides-de-camp send instructions to the different regiments. It was also our responsibility to deal with any problems that were bound to arise, and in particular it was more necessary than ever to ensure constant discipline.

Up to Newtown where the army was gathering, our only goal was to cover as much ground as possible. We were still too far from the enemy to have other precautions to take, aside from those necessary for provisions and discipline. From Newtown, General de Rochambeau decided to gather his forces to head toward the Hudson in tighter lines. On the evening of June 30th, he received a letter from General Washington asking him to hurry the march of the first division and the legion from Lauzun. General Washington had started the campaign on June 26th. Learning that the English General Clinton had divided his army into several corps dispersed around New York, he decided to attack the English part near Westchester, requiring support from French troops.

This surprise attack failed, but led to General Clinton's refusal to leave New York, even though he had received orders to send troops to Maryland, Pennsylvania, and Jersey to push General Washington east of the Hudson. This action also prompted Lord Cornwallis to retreat toward the opening of the Chesapeake Bay.

On July 6th, our army joined the American troops in Phillipsburg, where they stayed for a week. On the 18th, an aide-de-camp, Dumas, left in avant-garde to New York to prepare the reconnaissance that the two generals planned to do in person, to focus Clinton's attention and leave no doubt as to the allied generals' intentions. On the evening of the 21st, we left for the operation, with about 5,000 men total and two campaign batteries. The two generals did their reconnaissance together under enemy fire, also sending their aides-de-camp to find out about the English defenses. These operations lasted two days and allowed the capture of about thirty prisoners. On the return, General de Rochambeau kept telling us about the American troops, who were poorly dressed, poorly paid, made up of old men, children, and African Americans, but who marched as well on the roads as under fire.

The army stayed quietly in the camp at Williamsburg until August 14th. On that day, a message brought to Newport by the *Concorde* informed us that Admiral de Grasse had left on August 3rd for Chesapeake Bay with his fleet of 26 vessels, 3,500 men, and 1,200,000 livres. The situation was turning in our favor. The English general Clinton had locked himself into New York with 12,000 men, and could not attack us without weakening his defenses.

As for Cornwallis, he had been skillfully forced by Lafayette to withdraw to Yorktown, convinced that we had no artillery to expel him and that he would receive the necessary relief by sea. On August 15th, no one was yet certain of our destination, and we prepared several routes toward Kingsbridge and Staten Island, but also Philadelphia.

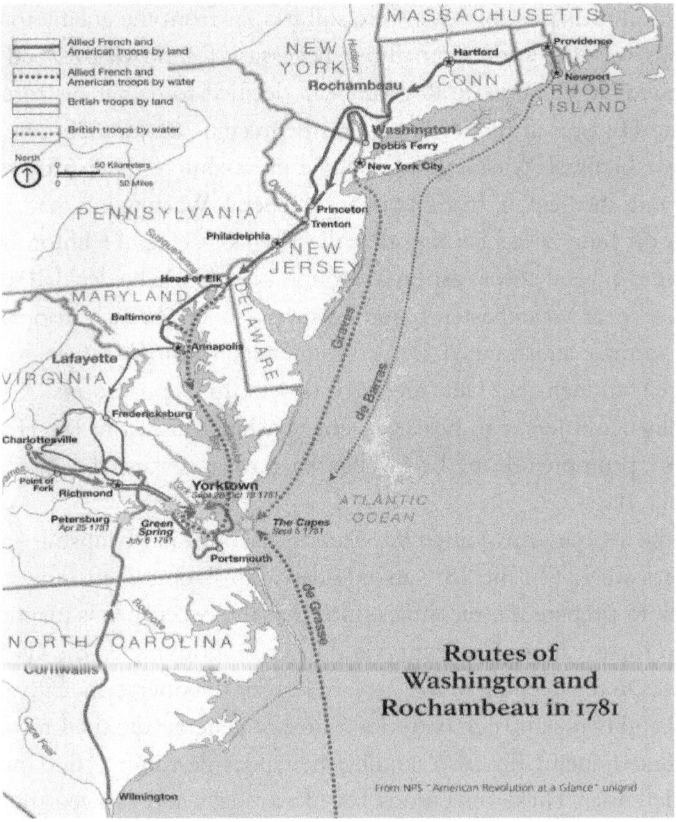

Routes of the American, French, and British troops in 1781.

After these maneuvers, the British were perplexed and suspicious. Fearing a ruse and convinced that our intention was to attack New York, they would never risk weakening their defenses to attack us. However, it was becoming clear to the allied command that to attack New York, especially without the support of Admiral de Grasse's fleet, would be suicidal, and that attacking Yorktown was the only possible action. Attacking Cornwallis had always been Comte de Rochambeau's idea.

I would like to emphasize here the skill of the two generals, true strategists, who maneuvered their troops in such a way that even our officers were convinced we were going to attack New York. I even wonder if General Washington's letter to Lafayette, ordering him to join him to attack New York, which had been intercepted by the English, had not been a decoy.

On August 19th, the two armies were clearly heading south. On September 3rd, we arrived in Philadelphia, where we showed off the troops in front of the population and especially the members of the Continental Congress. They were all impressed by our army. A good general, the Count de Rochambeau had not left his men idle during the winter. They were constantly occupied with work and exercise, and order and discipline reigned. As a result, the troops were ready to undertake a march of several hundred leagues and to maneuver perfectly.

Philadelphia was without a doubt the most beautiful city in America. The streets were wide and well laid out, with sidewalks on both sides for pedestrians. There were a large number of well-stocked boutiques and the city was very lively, with a population of nearly 40,000. In Market Street, there were two huge brick halls, one of them dedicated to butchers. The port, however, did not have much character, consisting of a dreary two-mile long pier. Like in all American cities, in Philadelphia there were several churches and a considerable college, which they call a university.[56]

The more we advanced south, I saw the people change; I saw less and less the biblical simplicity of the northern farmers, who worked their land and lived humbly with their families with one or two slaves at most. Here, the properties were much larger, the homes spacious, consisting of several buildings. The vast fields were cultivated by slaves that greedy Europeans purchased in Africa from Arab traffickers and sold at premium prices. The furniture was made from the most precious woods and the rarest marble, skillfully worked. The carriages, elegant and light, were pulled by steeds driven by magnificently dressed slaves.

---

[56] Balch, T. *op. cit.* p. 155.

The city of Annapolis was particularly rich with its beautiful buildings. The luxury of the women there exceeded that of our provinces, and a French hairdresser here was an important man.[57] Between the North and the South, I was in two different countries.

In Philadelphia we received the news that the fleets of the English admirals Hood and Graves were heading to Chesapeake Bay. As we did not know the position of Admiral de Grasse, this information concerned us. The troops nevertheless continued their route toward the city of Yorktown.

When we arrived in Chester on September 5th, we saw General Washington on the shoreline waving his hat in demonstrations of joy, behavior that was unusual for him. He had just heard from Baltimore that Admiral de Grasse had entered Chesapeake Bay with twenty-eight ships of the line and the 3,900 men of the Marquis de Saint-Simon, and that they had already reached land and joined Lafayette.[58] Monsieur de Barras had also joined Admiral de Grasse's fleet, bringing our siege artillery.

On September 10th, the Count de Rochambeau went to General Washington's home in Mount Vernon, on the road between Philadelphia and Yorktown. Mount Vernon was a large and beautiful residence, accessed by a vast esplanade. It was framed on each side by common ground and on the other side gave on to the Potomac River. The house, like the grounds, was perfectly maintained by numerous Black slaves. Our King Louis XVI had forbidden the possession of slaves in France since 1776, but it was still allowed here.

On September 18th, the two generals met with Admiral de Grasse on his ship. They learned that General Cornwallis had facilitated the task of the allied armies by entrenching himself on the Yorktown peninsula. He did not know that the French fleet was off the bay and thought that General Clinton would send him reinforcements or ships to transport his army; Yorktown was the perfect place for one of these solutions.

---

57 Abbé Robin, *op. cit*

58 Balch, T. op. cit. p. 157.

THE SIEGE OF YORKTOWN

On September 28th, the two armies arrived within sight of Yorktown. Since our departure from Newport on June 9th, the troops had covered more than 250 leagues[59] and set up thirty-nine camps.[60] For each step, we had to decide the marching orders for each of the regiments, determine which would be the front and rear guard, plan for the camps on arrival, and distribute the instructions making sure that they had been received and understood.

I had the surprise and great joy to find my older brother, Louis-Antoine, who had arrived a few days before us with the Marquis de Saint-Simon. He had embarked with his army of 3,900 men on the ships of Admiral de Grasse. Louis-Antoine, a major of the Touraine regiment with the rank of lieutenant colonel, was the major general of the army commanded by the Marquis de Saint-Simon.

The Marquis, on his arrival, had placed his troops under the order of the American army commanded by Lafayette while waiting the arrival of the Count de Rochambeau. We were able to take advantage of the woods, the curtains of trees, and the swamps to be within pistol range of the enemy, without losing one man. The American army had to build bridges over the marshes to reach their positions. The British were then surrounded on land by the French and American armies and on the sea by the French fleet. Clinton made a last attempt to liberate the bay, but the combined fleets of Admiral de Grasse and Count de Barras pushed back the English attack.

The French forces included the 6,000 men of General de Rochambeau as well as the 3,900 soldiers brought from Santo Domingo by the Marquis de Saint-Simon. The Americans were 11,000, including 3,000 militiamen. The two armies had to establish their camp. We gave precedence to our allies who decided to position themselves on the right, downriver.

Our artillery was very powerful, composed of sixty-four cannons, mainly Gribeauval, articulated in nine batteries. The Americans had sixty cannons. Our artillery was remarkable for the experience of our artillerymen and the quality of our cannons—particularly efficient and certainly the best in existence. Its role was essential, destroying fortifications, demoralizing the besieged, and especially protecting the diggers who dug the trenches parallel to the English fortifications in such a way as to minimize the distance to cross at the moment of attack. The experience of the French with sieges proved very useful in Yorktown.

---

[59] About 1,200 kilometers, or 750 miles.

[60] Abbé Robin, *op. cit.* p. 192.

Siege of Yorktown by Auguste Couder. Battle Gallery, Château de Versailles. François-Louis is wearing a tricorne hat and looking at General Washington.

Cornwallis had not waited for the arrival of the allied armies to fortify the city. Ten redoubts total were connected by trenches and filled with artillery. He had also fortified Gloucester Point, located on the other bank of the Chesapeake, a city he was in contact with thanks to two frigates, and whose defense he entrusted to Tarleton. This point was crucial, because it was through here that passed the provisions of the British. The French navy landed troops there on October 3rd, forcing Tarleton to retreat. The British lines of communication were now cut off. American and French methodically settled in, hardly bothered by English artillery fire.

From the beginning of the siege, I kept a journal of operations that I am including at the end of this memoire. Here I will simply evoke the main events of the siege.[61]

The rain that had fallen in the area had soaked the earth and we had no difficulty with the digging. On October 6th, the siege artillery had almost all arrived with the fascines, the gabions, and the racks prepared, and the location of the trench reconnoitered. The Count de Rochambeau gave the order to open fire the same night.

The work needed for the grand attack and to establish the batteries was done at night so the enemy could not see us. The goal was to build a trench that was wide enough, as close as possible to the fortifications as to group the soldiers there at the moment of the final attack. Two parallel trenches were therefore dug.

On October 9th, the French fired upon an enemy frigate, the *Charon*, a vessel of fifty cannons, attempting to go upriver. It took cover behind the city but too late, and it was hit and burned before sinking. Cornwallis had only one frigate left, which, threatened on all sides, could not leave its anchorage.

The Americans opened fire at the same time we did. The shelling intensified, and on the 11th, the British artillery was reduced to silence. Deserters told us that Lord Cornwallis had been surprised by the violence of the attack. His troops were disconcerted by this, because their general had assured them that the assailants were nothing to fear, despite their number, because they had no cannons.

The allies dug the second parallel a small rifle range away. For this to be able to stretch to the river, it was necessary to take over two redoubts that were on the way, and for that wait until the abatis and palisades were sufficiently destroyed to do this.

A night attack was planned on October 14th to take the redoubts 9 and 10, which were on the path of the second parallel. Under the orders of Baron de Vioménil, I was to command a detachment of the regiment of Soissonnais for the attack of redoubt n° 9,[62] while Lafayette attacked number 10. The taking of this redoubt would cost us forty-six men and sixty-two wounded. My services for this siege were rewarded with a pension of 800 livres.[63]

Two days later, the British, disguised as Americans[64] and taking advantage of certain French officers letting down their guard, managed to get close to a

---

[61] de Menonville, F-L. (1881) *Journal du siège de Yorktown.* Magazine of American History.

[62] Gardiner, A. *The Society of Cincinnati.* p. 96.

[63] Bodinier, G. *Dictionnaire des officiers généraux.*

[64] Abbé Robin, *op. cit.*

redoubt and tried to "spike" certain pieces of artillery,[65] but without success. A few hours later, the pieces were able to be fired once again.

With allied artillery raining down on the city, Cornwallis realized he could not hold out, and during the night of October 16th–17th, decided to evacuate to Gloucester; but a storm and artillery fire interrupted the attempt. Dysentery had ravaged his men, already demoralized and malnourished. With no news of a relief fleet, and after trying to delay with a cease-fire, Cornwallis surrendered. The British asked what kind of surrender they would be granted and George Washington told them "that of Charles-Town," reminding them of a victory for which they had made the Americans pay dearly. The British had refused the Americans "the honors of war," not allowing them to march with flags unfurled while playing the national anthem; they would endure the same humiliation.

Therefore, on the afternoon of October 19th, the British army filed between the French and American armies, in silence with their flags lowered. Cornwallis announced he was ill, doubtless to avoid attending this surrender that marked the ruin of many years of effort on his part; one of his subordinates carried his sword. He offered it to the Comte de Rochambeau, but the Frenchmen directed the officer to General Washington, and the British laid down their arms. There were 12,000 muskets, more than 2,000 pieces of artillery, and several cannons and munitions of war, which would now be used against them.

It was then I saw the hatred and contempt of the British toward the Americans. Despite their destitution, their torn and ragged clothes, after months of cold and hunger the Americans had fought like professionals with admirable courage. They greatly deserved this victory. For the British in their perfect uniforms, it was difficult for them to hide their shame at being defeated by such an army.

I later learned that the artist Van Blarenberghe had prepared a painting for the king, depicting the surrender as described by Louis-Alexandre Berthier. I knew Berthier well because he served under the Comte de Rochambeau. I never saw the painting, but I am told that in it I am depicted on horseback.[66]

---

65 "Spiking" a weapon was neutralizing it by sticking a nail in the touch hole through which the wick ignites the powder.

66 In fact, a study by Christopher Bryant formally identifies François-Louis in this painting from his uniform of adjutant general and the cross of Saint-Louis worn by the rider.
Bryant, C. *The Siege and Surrender of Yorktown, Van Blarenberghe's Paintings and the Metz Regiment of artillery.*

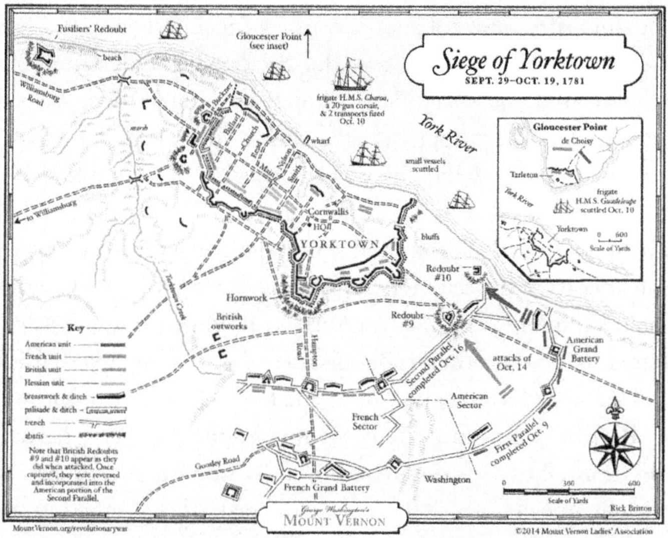

Map of the siege of Yorktown published by George Washington's Mount Vernon.

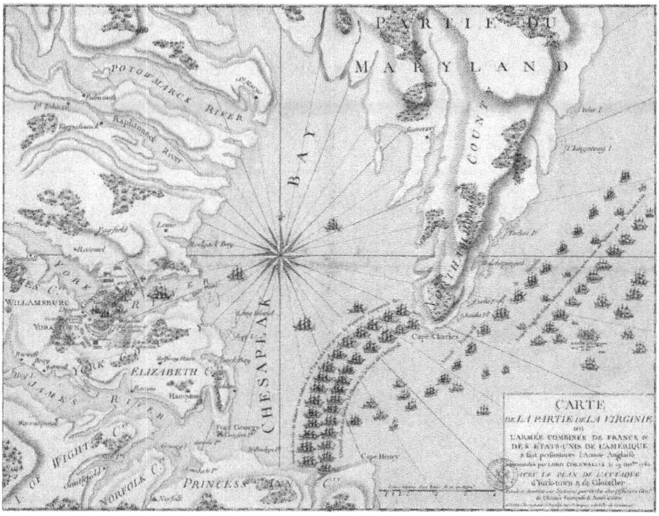

General plan of the siege of Yorktown.

Although no one knew it at the time, the American War of Independence was essentially over after the surrender of Yorktown. The wisdom and military science of the Comte de Rochambeau and General Washington had made this victory possible, with remarkably little loss of life. The capture of Yorktown cost the death of only fifty-nine soldiers and one officer. We counted less than 200 wounded—many fewer than the losses due to scurvy and dysentery!

Immediately after the surrender, Admiral de Grasse made his preparations for departure. He had the troops of Saint-Simon embark, and, on November 4th, sailed for the West Indies, which he was responsible for defending.

The troops dispersed to their winter quarters. The Virginia militia returned to the South, and General Washington sent Lafayette with troops from Maryland and Pennsylvania to reinforce General Greene's army to take Charlestown. My dear Baron de Vioménil was able to return to France, where he was needed for personal business; his brother took his place. I accompanied the Comte de Rochambeau and the Chevalier de Chastellux to our headquarters at Williamsburg where the Bourbonnais and Deux-Ponts regiments were garrisoned. The Chevalier de Chastellux took command of a division in December 1781 and the Comte de Rochambeau entrusted me with the task of replacing him in his duties as major general.[67] This was a recognition of my qualities in which I took some pride.[68]

The Siege of Yorktown by Louis-Nicolas van Blarenberghe, 1786. Gouache on panel, 23 1/2 by 37 inches. Private collection, U.S.A.

[67] Gardiner, A. *op. cit.* p. 96.

[68] Letter of the Comte de Rochambeau to Ménonville announcing his promotion to the rank of major general.

Detail from The Siege of Yorktown by Louis-Nicolas van Blarenberghe.

Detail from The Siege of Yorktown by Louis-Nicolas van Blarenberghe. Lieutenant Colonel François-Louis Thibault de Ménonville, Adjutant Major General.

The general positioned his troops so as to bring relief to the provinces most threatened by the enemy. In addition to threatening New York, he prevented the British from withdrawing forces there to fight us in the West Indies. The respective positions prevented both of them from taking any decisive action; the British were entrenched in New York and Charleston and could not get out, nor was it possible to attack these cities without suffering excessive losses. This winter period allowed the American army to equip itself. General Washington now had 12,000 soldiers of which he had about 8,000 drill before us. We were impressed by the transformation of this army, in its new and well-maintained uniforms, as disciplined and skilled at maneuvering as our own troops.[69]

The British approached the American Congress to sign a treaty between their two countries, excluding France. Congress refused this proposal twice, demanding the presence of their French ally. In the autumn of 1782, serious negotiations began between the plenipotentiaries of the three nations, and the preliminaries of a peace treaty were signed in Paris on November 30th, 1782. The final treaty was signed on September 3rd, 1783, and ratified by the American Congress on January 14th, 1784.

Starting in November 1782, the troops headed for Boston to board the ships of Monsieur de Vaudreuil's fleet. During my stay in Newport, I saw again many of the Americans with whom I had spent an earlier winter; it was undoubtedly the city where the French received the most appreciation and friendship.

The Comte de Rochambeau embarked in November, leaving the Baron de Vioménil, who had returned from France, in command of the army. Boarding of the ships began on December 7th, and the fleet left Boston on the 25th for Puerto Cabello, New Spain,[70] and then for Santo Domingo on the cape, where I found it painful to see the cruelty with which the slaves were treated on that island.

On the return trip, I reflected on the preceding months. It was similar to our alliance with Austria, when we returned as victors but without having gained anything on a material level for our involvement. What's more, the Americans would soon renew their ties with the British, with whom they were cônnected by blood and by language. We had helped friends, but had forgotten our own people, who were sinking deeper into misery. I did not expect to find France in better condition than when I left it. Our king had once said, "Only France wages war for an idea." Wasn't this generous vision in danger of backfiring on France?

---

[69] Clermont-Crèvecoeur. *op. cit.* p. 78.
[70] Venezuela today.

We finally arrived in Brest on June 17th, 1783. We just had to organize the landing of the troops and determine the destination and the marching orders for the regiments to their camps.

RETURN TO THE VOSGES

As soon as I could, I returned to the Vosges and to Charlotte de Martimprey de Choisimont, whom I would marry on September 29th, 1783. On June 18th, 1784, our son François-Charles was born in Deneuvre. On the 1st of December, 1784, I was appointed a brigadier, the last rank before major general.[71]

When I returned to the Vosges after more than three years' absence, calm reigned, as in most French provinces. Everyone affirmed their affection and loyalty to the king. The people of the Vosges were simple and good, mostly farmers, as there was little industry or factories. Many families lived in great poverty, with the priests trying to help as they could with their limited means. There were few strongholds, most having been destroyed by order of Louis XIV. The nobility were few, not very influential, and often close to the people.

The clergy, small in number, were modest and poor, living among the people. An exception, however, was Bishop Chaumont de La Galaisière, who was arrogant, lived lavishly, was surrounded by dandies for vicars general, and was quite harsh with his priests.[72] The abbeys were numerous and rich, with those of the women the most opulent. The abbess of the noble ladies of Remiremont, a princess by blood, only travelled in a carriage with six horses and her stipend was 300,000 livres per year.

There were several Masonic lodges, and the largest one was the Incorruptible Friends, founded at Saint-Dié in 1785, but they had little impact.

The situation in the rest of France was already bad when I went to America. Louis XVI tried to straighten out the country at the beginning of his reign by calling on Turgot, Comptroller-General of Finances, who initiated some essential reforms. Yielding to the hostility of those affected by these reforms, the king dismissed his minister in May 1776. It was reported to me that before leaving office Turgot said to the king, "Never forget, sire, that it was weakness that put Charles I's head on the block!" Obviously, this warning was not heeded!

Louis XV had succeeded in subduing the parliaments which had been in rebellion against royal authority since the death of Louis XIV. Louis XVI made the mistake of recalling them as soon as he came to power. They immediately resumed their unrest and did everything they could to undermine royal authority. The king then appointed Necker Minister of Finance in 1776. An opportunistic banker from Geneva, he understood that it was easier to borrow than to tax. It

---

[71] In the Ancien Régime, the intermediate rank between colonel and major general.
[72] Bouvier, F. *Les Vosges pendant la Révolution.*

was a loan, moreover, that financed our campaign in America. Necker, however, eventually realized that it was impossible to keep borrowing and that Turgot was right. It was necessary to distribute the collection of taxes differently. To this end, he proposed the establishment of provincial assemblies whose mission would be to define the distribution. This proposal failed, because the Parliament of Paris refused to cede part of its powers to provincial assemblies. In the meantime, Necker had made a powerful enemy. He also understood that a reduction in expenditures was essential. He managed to limit those of the ministries but not the royal court. To justify his actions, he decided to publish a report to the king detailing the revenue and expenditures of the treasury. The report, which became public, showed that the annual spending of the court amounted to a staggering 33 million livres. This information provoked harsh criticism and further discredited the regime. Necker had to resign.

In the same period, the Parliament of Paris went to war against the lettres de cachet, letters bearing the seal of the sovereign and allowing him to have whomever he wanted arrested or exiled without justification. I recognize that this right was, in principle, from another era, but in practice this measure was mainly used at the request of families anxious to preserve their reputation by requesting the imprisonment of one of their own. This justice of the king was often less severe than civil justice, which easily sent the guilty to the executioner or the galleys. It was thanks to the king's justice that Mirabeau, condemned to death for having fled to Holland with the wife of the Marquis de Monnier, was imprisoned in Vincennes and thus escaped the executioner.

Even before my departure for America, partly because of Parliament and the newspapers, Paris was becoming hostile to the king and the court. The queen was becoming less and less popular, and her frivolity and behavior gave rise to the worst gossip. She was nicknamed "the Austrian," and her expenses earned her the nickname "Madame Deficit." But at Versailles, no one was concerned, or wanted to be concerned, with the situation.

Charles de Calonne was appointed Controller General of Finance in November 1783. The financial situation was becoming catastrophic, with the debt reaching three billion livres. Calonne found no other solution than to stay the course until the loan was no longer covered. Once again, Turgot's proposals were revisited and Calonne proposed introducing a tax to be paid by all landowners and proportional to income.

Louis XVI supported this idea and set up a committee of prominent people to approve it. This committee, composed of privileged people who did not want to pay this tax, hastened to reject the proposal. The king, as he often did, retreated before the show of strength and gave up; his minister was dismissed in April 1787. Loménie de Brienne succeeded him, and could only take up the project of his predecessor which called for a tax on everyone, including the privileged class, who had been exempt until then. To pass this measure, the king decided to present a royal edict. He went personally to Parliament and had a text read there imposing his will on the magistrates. Unthinkably, Parliament declared the recording of the edict "null and void" the next day.

Brienne responded by exiling the parliamentarians to Troyes, but they were joyfully welcomed there. In Paris, demonstrations were organized against the king and especially the queen, whom the masses referred to only as "Madame Deficit." The twelve provincial parliaments joined the protest. Brienne had to renounce the tax plan and, to calm the turmoil, promised to convene the assembly in 1792. The return of Parliament to Paris was also authorized; Brienne hoped to raise another loan, but Parliament refused. A second royal edict was then decided.

The Duke of Orleans declared "It's illegal."

The king replied, "It's legal because I want it," but stammered and spoke without conviction.

Opposition to royal power was organized around Parliament, bringing together the privileged who wished to retain their privileges and the patriots who demanded freedom.

To my great surprise, this protest movement enjoyed the support of the Duke of Orleans, the king's cousin. It had already occurred in France that princes opposed the king, but this was the first time, to my knowledge, that one of them sought to destroy the monarchy. His hatred of the king and queen was fierce and he lost no opportunity to harm their position. Louis-Philippe d'Orléans, Grand Master of Freemasonry in France since 1773, lived for pleasure and was extremely rich but always looking for money. He lived in the Palais Royal, which he transformed into the headquarters of the protesters. The king exiled the duke to his lands, but this was short-lived and mostly strengthened the unrest.

I was devastated. When the king did nothing, it was a mistake, and when he did take action, it was still a mistake! Every good decision was inevitably reversed. He tried to diminish the influence of the parliaments in May 1788 by creating new jurisdictions, but the parliaments, supported by all the discontented, entered into open revolt. This was particularly vivid in Grenoble, where the crowd forced the parliamentarians into the courthouse and threw tiles at the army that had come to remove them. In July, the provincial states of Dauphiné, composed of representatives of the Third Estate but also many nobles and clergy (165 and 50 respectively) demanded a meeting of the Estates General. Brienne yielded again, and set the meeting for May 1st, 1789.

It remained to establish the composition and mode of operation of these Estates General. The parliaments, finally realizing the danger, asked that the structure of 1614 be preserved: three chambers that would vote separately. The political clubs, which had once again been authorized, stirred up public opinion and demanded a doubling of the number of deputies of the Third Estate, which Parliament finally accepted.

Surprisingly, the king's council ratified the doubling of the number of deputies of the Third Estate in December 1788, but left it to the orders to decide the vote by order or by head. The king and his council were convinced that this summons would be purely formal and inconsequential.

Curiously, the announcement of the convening of the Estates General aggravated the unrest. Many riots broke out in the provinces, invariably ending in looting. In Paris, a riot caused by a wage cut was suppressed by the army. There were more than 300 dead and a thousand wounded. This was the only case of repression, as the king subsequently forbade the use of force.

In the midst of this bad news for our country, I received at least some good news for myself. A letter from General Rochambeau, dated Sept. 21st, 1788,[73] informed me that I had been appointed major general. He reminded me in the letter that I had already served this rank, as I had had this responsibility in America in December 1781, when the Chevalier de Chastellux, major general, had taken command of a division. This appointment was accompanied by an increase in my pensions to 6,200 livres.

---

[73] Letter of Comte de Rochambeau to François-Louis Thibaut de Ménonville, September 21st, 1788.

DEPUTY OF THE NOBILITY TO THE ESTATES GENERAL

A few months later, in March 1789, the electors who were to vote for the representatives to the Estates General were designated. There were to be thirty-six deputies for all of Lorraine, including eight for the bailiwick of Mirecourt, of which I was a member, and therefore two members representing the nobility. On March 31st, the electors met to designate the deputies. Despite my differences with the presidents, I was elected without difficulty. My nomination as major general had strongly played in my favor; the other deputy of the nobility, the Count de Toustain, was also a major general.

Elections did not go so well everywhere, especially in Nancy and Pont-à-Mousson, where voters were circumvented. In Nancy, the Chevalier de Boufflers spent 6,000 livres to provide dinner to the electors, which led to the saying that "he had won 200 turkeys with 50 chickens."

The list of grievances was drawn up jointly by the three orders. I had written my own report in which I prescribed the vote by head for questions relating to taxation and those common to the three orders, and the vote by order on the privileges and special interests of the nobility.[74] With the clergy, we had at least partially surrendered our monetary immunities. We asked for the creation of a printing press and a barracks for Mirecourt. In their report, the deputies of the Third Estate assured the king of their loyalty and fidelity, offering him "their blood, their life and the little property they have left." The grievances were similar to those that would be found in the rest of the country: restoration of the provincial states, regular meetings of the Estates General, accessibility to all jobs, abolition of lettres de cachet, freedom of the press, and a better distribution of taxes.

Once this list of grievances had been written, I prepared to go to Versailles where the Estates General were to meet. The easiest way to get to Paris was the turgotine, a new carriage that had replaced the smaller coaches. It took less than four days to get from Nancy to Paris. Of course, each one carried up to fourteen people, but a private carriage, which had to use the same horses for the duration of the journey, took much longer and was more expensive.

When I arrived at Versailles, I learned that lodging was prepared for me at the Grand Commun, a vast building adjoining the castle that housed the king's

---

[74] Bodinier, G. *Dictionnaire des officiers*, *op. cit.*

officers and the staff working in the castle. Only twelve deputies had that privilege. I lodged at number 54, near the stairway of the queen's office.[75]

Versailles offered a great variety of accommodations, and everyone could find what suited them depending on their means. For our meals, we had the choice between meals with the individuals who boarded us, inns, or one of the many caterers. The minister ordered the creation of tables d'hôtes, tables with fixed menus that could accommodate more than half of the deputies. The best was at La Vènerie, the most expensive at Au Chenil and Les Menus Plaisirs. There were also more modest ones. In any event, the allowance paid to the deputies, which was 540 livres a month, was more than enough to cover my expenses.

At first glance, Versailles was a beautiful city, with very wide streets lit by oil lanterns and sidewalks protected by posts lining them, which allowed safe walking in all weather. However, these streets were cluttered with rubbish, and ditches where garbage and dead animals were dumped were rarely cleaned. Stalls were set up everywhere with the most diverse businesses: booksellers, drinking establishments, and sellers of fruits, vegetables, and other foodstuffs. These shops were found in all the streets up to the courtyards, vestibules and even the corridors of the palace.[76] There were also very pleasant cabarets where you could enjoy coffee, which was very fashionable here. The police had a lot to do: arguments broke out all the time, and a foundation of crude and violent morals was found among the people as well as among the officers and people of quality. Every day, quantities of objects disappeared, homes were broken into, and thefts were committed—even in the royal apartments or at mass.

I had some surprises as soon as I arrived. First of all, to find that the deputies of the Third Estate were neither workers nor peasants, but for the most part educated petty bourgeois, very often lawyers, frequenting the same salons, clubs, or lodges as the educated elites. Then I learned that we would have uniforms; black for the Third Estate and colored with a dominant accent of red for the clergy and nobility. The ceremonial costumes were identical, except for those of the nobles, which had green shoes instead of red. The deputies of the Third Estate expressed their disapproval of having a black coat imposed upon them, which contrasted with the garb of the other two orders.

---

[75] Brette, A. *Recueil de documents relatifs à la convocation des États généraux de 1789.* Tome 2.
[76] Hindie Lemay, E. *La vie quotidienne des députés aux États Généraux.*

DEPUTY OF THE NOBILITY TO THE ESTATES GENERAL

Clothing of the deputies of the three orders at the Estates General. Gallica. Unidentified print.

I found among the deputies thirteen "Americans," French citizens given that honor because they fought in support of the revolution. Eleven of them represented the nobility: La Fayette, Charles and Alexandre de Lameth, the Marquis de Saint-Simon, the Duke of Noailles, Biron, Victor de Broglie, General Custine, the Duke of Castries, and Mirabeau the younger. The latter, unlike his brother, was a legitimist, and his stoutness and affection for wine had earned him the nickname "Mirabeau-Barrel."

They were mostly old acquaintances, and even if some of them at times made me feel the different circumstances of our birth, we had a common past that brought us closer together. We had many opportunities to meet and discuss matters, although we did not always agree on how to ensure the future of our country.

Clothing of the deputies of the nobility to the Estates General. Gallica. Clothing, insignia, cards, medals of deputies, 1789–1898 Edmond Launay.

At one of these meetings, the Marquis de Saint-Simon informed me of a letter he had received from General Washington, in which the latter asked him to congratulate me and convey his greetings. A little surprised by this unexpected solicitude, I spoke of it to Lafayette, who told me that it came from him. Indeed, the general needing to be in the good graces of Marquis de Saint-Simon, he had recommended that the general compliment me to him, because the Marquis liked to hear good things said of those who were close to him.

The opening of the Estates General was scheduled for May 5th. There was a procession on the 4th that led us from the Church of Notre Dame to the Church of Saint Louis. Parisians came in large numbers to witness this procession, so many that the French and Swiss guards had difficulty containing the crowd. It was magnificent. The wide streets of Versailles were adorned with splendid tapestries, the king was escorted by officers in full dress and members of his family, and the whole royal court was present.

The opening session took place the next day. We were gathered in a large and beautiful room of the Hotel des Menus Plaisirs, and many pretty women of the royal court came for the session. The king arrived accompanied by his two brothers, and gave a speech much appreciated by all. Necker then spoke and bored us for three hours. As Mirabeau was about to deliver his speech, the king suddenly decided to adjourn the meeting, refusing to let him talk. It was not planned for him to make formal remarks, but it was not a good way to inspire the goodwill of an important member of the Third Estate.

Starting the next day, the Third Estate refused to meet separately, and the deputies of the clergy decided by a narrow majority to join them. As for us, we refused. Father Sieyès, a member of the clergy but elected by the Third Estate, proposed that the chamber of the Third Estate become a national assembly. This proposal, which was in fact a decision, was communicated to the king. In response to this injunction, Louis XVI closed the doors of the meeting room of the Third Estate. They then found a gymnasium where tennis[77] was played and gathered there. It was there they took an oath to remain united until the constitution of the kingdom was established, which was later known as the "Tennis Court Oath."

The king gathered deputies of all orders on June 23rd to inform them of the concessions he was granting. He ended his speech by asking the members to return to their seats by order. The deputies of the Third Estate refused to move. Two days later, forty-seven deputies of the nobility, led by the Duke of Orleans, joined the Third Estate in their turn. Heavens, how repugnant this man was! He was described as morally bankrupt, without talent, leading a villainous life, greedy for money, with all the vices that made one hate crime.[78] Another member deeply disgusted me: Mirabeau. Adorned with a huge hairdo, he was ugliness personified—fat, sloppy, his face marked by smallpox. Rivarol said he was "the

---

[77] Predecessor of modern tennis.

[78] Quetel, C. *Crois ou meurs.* Éditions Taillandier.

man of the world who most resembles his reputation: he is awful." He was dangerous, because he knew how to speak to others and especially how to lead them where he wished; he addressed the senses and not the mind. The Marquis de Ferrières rightly said that he shaped men at will—that it was not a question of convincing, but rather that it was necessary to captivate, to subjugate[79] them. A proud man, he boasted and got drunk on his own words and persuasiveness. Not all the leaders were like him, but I felt the appearance of ambitious men, ready to do anything to ensure their domination and establish their power.

The king retreated again on June 27th, asking the clergy and the nobility to join the Third Estate. Of course, I agreed to obey this order, but like the Marquis de Saint-Simon,[80] I refused to participate in the head count vote until I had been authorized to do so by my constituents.[81]

Joseph-Désiré Court. June 23rd, 1789, at Versailles, Mirabeau. Rouen, Museum of Fine Arts.

[79] *Ibid.*

[80] Hindie Lemay, E. *Dictionnaire des Constituants.*

[81] *Bodinier, G. Dictionnaire des officiers, op. cit.*

Even if the information we had at Versailles was contradictory, it became more and more evident that the situation in Paris was becoming worrisome. Bread had never been so expensive, and the number of unemployed and hungry people was constantly growing.

Demonstrations were organized on the Champs-Élysées, and demonstrators attacked men on horseback, who were forced to charge. But at Versailles, no decisions were made. The Baron de Besenval, commander of the troops, was left on his own. To avoid further clashes with demonstrators, he withdrew the troops from Paris, leaving the field open to crowds and riots.

News spread like wildfire at Versailles on the evening of July 14th. Riots broke out in Paris: a few thousand people had succeeded in seizing rifles and cannons at the Invalides while the regiments withdrew from the Champ de Mars. Fortunately, there was no ammunition. The rioters then went to the Bastille, an impregnable fortress defended by cannons and ditches. The frantic warden tried to reason with the rioters, and finally opened the doors without any cannons being used. Those who knew the area claimed that it would have sufficed to leave the doors shut to prevent the invasion… and looting.

Paris celebrated this victory with drinking and the conquerors of the Bastille, happily drunk, roamed the city accompanied by prostitutes and the sans-culottes—supporters of the Revolution, called thus as they wore trousers instead of the breeches of the aristocracy. They were beginning to multiply.[82] To thank the warden for not having used the cannons, the rioters paraded the streets of Paris with his head and that of the provost of the merchants on a spike.

Meanwhile, the king remained indecisive and took no action. His advisors had different opinions and he did not know whom to follow, so did nothing. He was even forced to go to Paris, accompanied by thirty-two deputies and an escort that looked more like a troop of looters than guards. Disorder and agitation pervaded Paris and spread to the provinces, where it turned into a revolt against rights and fees.

In the National Assembly, the delegates wanted to put an end to this anarchy, but the only troops available were those of the king and it was impossible to use them. To appease the rioters, they offered concessions. On the night of August 4th–5th, we decided to abolish feudal rights against buyback. Tithes were also abolished and equality before taxation proclaimed. The result was that no one paid taxes from that moment on.

---

[82] Chateaubriand, F-R. *Mémoires d'outre-tombe.*

The state coffers were empty, and riots, mostly provoked by hunger, occurred in Paris and in the provinces. The Versailles nobility increasingly opted for exile, leaving their staff, now without jobs or income, to join the ranks of the hungry in Paris. They were joined by the workers and luxury craftsmen who no longer had customers. These issues did not disturb the members of the National Assembly, who were working on drafting the Declaration of the Rights of Man and of the Citizen and a constitution. Two worlds coexisted: ours, of speeches and theories, and the other, of revolt, famine, and suffering, which our assembly haughtily ignored.

Several thousand women from the marketplace marched on Versailles at the beginning of October. Among them was an agitator, Fournier. He was a brutal, vain, and hateful man, and he later claimed that he himself had instigated the march. Other demonstrators joined them. They remained in Versailles all night, and in the morning a group managed to break into the palace. The queen, warned in extremis by a bleeding servant, only just had time to take refuge with her husband.

When order was finally restored, the royal family had to appear on the balcony. The crowd shouted "the king to Paris," and Louis XVI agreed to go. I was one of the delegates chosen to accompany him.[83] The procession would be preceded by cannons, straddled by thieves and prostitutes shouting obscene remarks. Then in the midst of a horde of all ages and genders followed the bodyguards on foot, each of their horses ridden by two or three drunken and scruffy lowlives.[84] We would come next, preceding the carriages of the king and his family, escorted by two soldiers' heads, the hair curled and powdered by a wig maker, stuck on halberds.

The meeting in the morning determined that we would accompany the king to Paris. We had to leave Versailles on October 19th, as soon as we were ready. This idea appealed to us, because, in the large room where we met, we had to endure the heavy and pestilential air exhaled by more than 3,000 people crammed together, to the point that many of us were sick.

One hundred and twenty delegates of the nobility resigned at that time. I hesitated for a long time whether to join them. I was disappointed by what was happening, but did I have the right to give up? Wasn't that an easy and somewhat

---

83 Parliamentary archives, 6 October 1789.

84 Chateaubriand, F-R. *op. cit.*

cowardly solution? It seemed to me that we should at least try to limit this debacle, so I decided to persevere.

I found accommodations in Paris at the Hôtel de Condé, at 19 rue Traversière Saint-Honoré.[85] This accommodation had the advantage of being close to the Tuileries and the hall Manège where we met. I had already been to Paris several times when we were staying in Versailles. This city offered a violent contrast between the beauty of its buildings and the dirt and stench of its streets. Walking was difficult because of the rubbish there, and dangerous because of the carriages that passed through without any concern for pedestrians.

On the other hand, the social life was even brighter than at Versailles. There were literary meetings, political societies, and shows everywhere, and the theaters were always full. The walks along the boulevard du Temple, the boulevard des Italians and the pathways of the Tuileries gardens were full of elegant women. A multitude of carriages crisscrossed the avenues. Famous courtiers, such as the Duke of Lauzun, a chaser of women, and the Baron de Bezenval, joined the tricolor flag and were now employed by the Revolution. After needing their vices, it now needed their blood.[86]

The functioning of the National Assembly was becoming increasingly chaotic. There were more spectators than delegates, and they had to arrive early to find a seat, as the galleries were so crowded. The sessions would begin with the reading of the minutes of the previous day, and then move on to the agreed-upon subject, unless there was a special motion. Everyone spoke and the debates were mostly stormy. Those in the galleries were noisy, applauding their champions. They were all gathered on the left side of the president, and forbade the right side from speaking, where moderates like me and the royalists were. The president rang his bell, usually without much success; members shouted from one side to the other.

The women from the marketplace knitted underwear during the sessions. When a conservative delegate made a proposal they did not like, they stood up and shouted "To the lantern!" an exclamation referring to hanging someone from the poles used for street lanterns. The debates could also be interrupted by petitioners who, armed with spikes, came to the bar to state their demands. Most often, they asked for bread. Some of us were insulted during the sessions and in the streets, received death threats, or were jostled at the exit of the hall. Others

---

85 Rue Molière today.

86 Chateaubriand, F-R. *op. cit.*

received threats from the Palais Royal. Father Maury had to put up a fight several times because they wanted to expel him from the gallery.

Alexandre Debelle.
Constituent Assembly, the demands of the people.
Album du Dauphiné.

I was appointed assistant to the Committee on the Navy in January 1790, but I resigned in March 1791 in protest; there was dissension within the committee and one member made an atrocious accusation against absent colleagues.[87]

I limited my contributions to subjects that I knew well, especially when I could enlighten the National Assembly with my knowledge of English laws. In December 1789, I proposed a draft decree on the mode of deliberation and the administration of the property of the communes. This bill was referred to the constitution committee for review, which was a way of rejecting my proposal.

I had to intervene to contradict Duport in June 1791. He showed great ignorance of English law in his speech on the right of pardon. I pleaded for the king to have this right.

---

[87] National Assembly parliamentary archives, March 1st, 1791.

I also spoke on the organization of judicial power, the rights of the accused to refuse jurors, and the composition of the jury for the National Superior Court. Without needing to speak at the bar, I also took a stand against the abolition of the nobility in June 1790 and signed a petition in favor of Catholicism as the state religion. Voting was by show of hands. One day I called for a roll call vote because I was convinced that the president had miscounted the result, and I was accused of insulting him. Some even demanded that I be imprisoned.

The National Constituent Assembly became increasingly totalitarian, exceeding its powers by creating the Research Committee to receive denunciations against the perpetrators of public misfortunes. Even Mirabeau was concerned about this development. One delegate proposed that votes should henceforth be taken by secret ballot and behind closed doors. His proposal was booed and rejected. The left side wanted terror to reign and be able to intimidate the opposition by identifying opponents through roll call votes.

Public opinion and the assembly were manipulated by the Jacobin Club.[88] Their leaders were sectarians for whom any means were acceptable for achieving their goals. They intended to exercise all the power without sharing it, and they were allowed to do all of this since they proclaimed themselves "liberators." Any opponent was declared an enemy of the people. They even went so far as to disrupt the meetings of other clubs to have them banned, on the pretext that there was disorder!

The newspapers were mostly Jacobin and often extreme, sometimes trashy. By far the worst of them was L'Ami du peuple ("The People's Friend"), whose sole editor was Jean-Paul Marat. He vilified, accused, denounced people and demanded punishments. He declared that the people had to kill 270,000 persons in order for the Revolution to progress. He was referring, of course, to the nobility. Fortunately, Charlotte Corday rid us of this bloodthirsty madman.

There was also Les Actes des Apôtres ("Acts of the Apostles"), a royalist-leaning newspaper that was a great success. The fight waged by Clermont-Tonnerre, Mirabeau-Barrel, the Vicomte de Ségur, and Rivarol reassured me. I could not abandon these brave men by resigning, as others had done and some had suggested to me.

The assembly's incompetence in matters of finance was glaring, so the situation of the treasury was catastrophic. Taxes were no longer coming in and

---

[88] Name derived from the convent of the Annunciation or the Jacobins in the rue Saint-Honoré. The term "Jacobin" refers to the Dominicans.

spending was increasing, especially to help the most disadvantaged and to subsidize bread prices to avoid grain riots.[89] Necker proposed a compulsory loan from the richest citizens. The assembly was appalled and refused on the grounds that it would be a violation of the right to property established in the Declaration of the Rights of Man and of the Citizen. It must be said that many members of the Third Estate were concerned by this measure! We would later see that our assembly did not always have the same qualms about respect for human rights.

In effect, the delegates proposed that the State appropriate the wealth of the Church. Obviously, for some, including Talleyrand, the clergy were neither men nor citizens covered by the Declaration of Human Rights, which guaranteed the right to property. Talleyrand thus betrayed his order and his robe. On November 2nd, 1789, the day of the vote on this provision, the crowd started gathering at 6:00 a.m., and the delegates who were members of the clergy were insulted and manhandled. Several were prevented from entering the session and participating in the vote. The confiscation was then voted. While waiting for the items to be available for sale, assignats were created. Then there was paper money. These creations would allow speculators to buy confiscated goods, with assignats constantly devalued. Even Necker tried unsuccessfully to limit this proliferation of artificial money, and finally resigned in September 1790.

The confiscation of property was followed by the reform of the organization of the Church. From July 1790, bishops and parish priests would be elected by taxpaying citizens, whether they were Catholic or not. The clergy would be paid by the state. The pope would wait until March 1791 to condemn the civil constitution of the clergy and the attacks on religion. He would also later condemn the oath of the elected prelates. The Church of France then split between jurors and those who refused to take the oath. This crisis provoked unrest and riots everywhere because, most often, the people recognized only the non-sworn as priests, especially in the Vendée. In Avignon, the population rose up to protest against the annexation decided by the assembly. The revolutionaries took revenge by massacring dozens of the pope's supporters who were detained in Paris.

Following a speech by Louis XVI to the assembly in February 1790, a civic oath was proposed, swearing loyalty to the nation, the law, and the king, who had accepted the Constitution. A great federative movement then developed throughout France, but extremists ruined this trend. Marat, now helped by

[89] Revolts due to the cost of wheat.

Camille Desmoulins, constantly called for murder. Robespierre, leader of the Jacobin Club, became more and more intransigent and began to exercise a real despotism of opinion. The Terror was beginning.

Faced with this extreme left, which was still a minority, a majority of delegates, including myself, joined the center-right. I congratulated myself on this move and also for deciding not to resign. Mirabeau's successors approached the court in an attempt to restore order. These men would not accept anarchy and wanted to preserve their gains by fighting any excesses.

In early November 1790, I was able to obtain a copy of Edmund Burke's book, which had just been published in London: *Reflections on the Revolution in France*. I remember reading this book during a trip to the Vosges.[90] I had crossed a devastated, unhappy, and desolate country. Our delegates were not aware of the evils that their carelessness and incompetence inflicted on the French people, whom they claimed to defend and protect. Mr. Burke's book was fascinating, and I shared many of the criticisms it made. His picture of the situation in France was strikingly realistic: "They have found their punishment in their success: laws overturned; tribunals subverted; industry without vigor; commerce expiring; taxes unpaid, yet the people impoverished; the Church pillaged with no benefit to the State".[91] Several aspects, however, seemed questionable to me, and as soon as I returned to Paris, I decided to share my remarks with the author. I wrote a letter to Sir Edmund, in which I pointed out some inaccuracies in his comments on the work of the assembly—minor confusions, but I feared that they might lead readers to question the otherwise excellent work as a whole. I was more concerned about the consequences of Mr. Burke's criticism regarding the future of our institutions. Created by what was now called the National Constituent Assembly, although they were revolutionary, those institutions remained relatively moderate.

The assembly had already gone far in its reforms, but the result was reasonable because the monarchy, although weakened, was preserved. Edmund Burke did not share this view and considered that a monarch who accepted being placed in the position imposed on him by the Constitution would never be up to his role.[92] Perhaps another king could return a more satisfactory balance.

One criticism of Mr. Burke's particularly offended me. He wrote of the assembly, "after a gang of assassins had driven away hundreds of their members,

---

[90] Letter of François-Louis to Edmund Burke.

[91] Burke, E. *Réflexions sur la révolution en France*. Les Belles lettres. p. 41.

[92] Burke, E. *op. cit.* p. 212.

while others, who held the same moderate principles as those who were driven away, but held on because they had more patience or better hope, are daily exposed to outrageous insults and murderous threats," referring to the departures of Messrs. Lally-Tollendal and Mounier, who were followed by many others. Mr. Burke felt that these members had done well to leave the House and criticized those who had remained, such as myself. I had, with a few close to me, carefully thought about what should be done. We concluded that these departures would weaken the defense of the king and our country. The departure of the moderates was indeed the dearest wish of the Jacobins and all the extremists, for they would then have a free hand. True courage was to stay and defend our country and our king as much as we could. It is true that more and more the crowd booed the moderates and went so far as to jostle them. The committee of inquiry was beginning a reign of terror. That is why I asked Edmund Burke not to mention my name in his reply to my letter.

I sent my letter on November 17th,[93] and Mr. Burke's reply, dated January 19th, 1791, responded to my remarks but also contained truths which I dared not formulate. He considered that "it would be better for him [the king] to cease living than to continue to be the passive instrument of tyranny and usurpation".[94] Mr. Burke also predicted that the king would be removed as soon as the revolutionaries could do so. I did not imagine then that it would be possible. His prediction was exactly two years and two days ahead.

A new incident occurred on April 18th, 1791. As the king went to his castle at Saint-Cloud for Easter, a crowd blocked his passage, insulting and threatening the royal family. Lafayette proposed to intervene, but Louis XVI replied, "I do not want blood to be shed for me." The king proved once again that he had no understanding of his role as monarch. He had not understood that his person did not matter, that a king was only a link in a long chain of men who had to ensure the survival of the monarchy. He should have known from the beginning of the troubles that firmness was the only way to avoid thousands of deaths and the end of the monarchy. Henry IV had been a king in the service of France and its people; since Louis XIV, it was France and the people who were in the service of the king. This mistake was fatal to our monarchy.

Finally, tired of the treatment he was receiving and after much hesitation and delay, Louis XVI decided to leave Paris on June 22nd. Again, the errors committed

---

93 *Annales historiques de la Révolution française.* 27ème année, n° 140, p. 217-227.

94 Burke, E. *Lettre à un membre de l'Assemblée nationale.* Mille et une Nuits, p. 36.

by those who were to ensure his protection, the inconsistencies of the king who got out of his carriage and showed himself, and bad luck and incompetence ruined this attempt. The king was arrested and brought back to Paris in a humiliating manner. He was now nothing. In the assembly, it was said that the king had not fled, but had been kidnapped. No one believed this pious lie, and the madmen of the Jacobin Club were only too happy to add fuel to the fire. All hope of a return to calm had vanished. The Jacobins, the press, and the people of Paris attacked Louis XVI and he was insulted, vilified, and ridiculed. The revolutionaries claimed that France no longer had a king. Nevertheless, a moderate action emerged, as many Jacobins left the club and created the group called the Feuillants, attempting to isolate the extreme left led by Robespierre.

The Baron of Vioménil was to accompany the king but had to leave his place to a governess of the Dauphin. I am convinced that, had he been present, things would have turned out differently.[95]

A demonstration calling for the proclamation of the republic was organized on the Champ de Mars. In anticipation of unrest, martial law was decreed by Bailly, the mayor of Paris, and Lafayette. The latter, on his arrival at the Champ de Mars, was pelted with stones. The insurgents refused to disperse, a shot was fired, and a shootout ensued. There were more than fifty dead.

It was decided to close the Cordeliers club. Named after a Franciscan monastery, this club was even more extreme than the Jacobins, stoking agitation and stirring up riots. Hébert was arrested while Marat was in hiding. Bailly, who had come to explain the situation to us, was congratulated. But the press went on a rampage against the city of Paris, calling its action "the Saint Bartholomew of the patriots." The Revolution was now split in two. The National Constituent Assembly was dissolved and replaced by the Legislative Assembly, and it set to work on a new constitution.

On September 8th, I protested against this constitution[96] which, unlike that which existed in England, entrusted most of the powers to a single national legislative assembly. There was no provision in case of conflict. The king could not dissolve the assembly and the assembly could not overthrow him. The king took an oath to the Constitution, even though many on the right and left thought it was doomed to failure. Robespierre had already begun to declare that the

---

[95] Bodinier, G. *Dictionnaire des officiers*, *op. cit.*

[96] Hindie Lemay, E. *Dictionnaire des constituants. op. cit.*

Revolution was not over, and the Jacobins wanted the monarchy to disappear. Camille Desmoulins compared it to a mountain of ice sitting on top of a volcano.

The new assembly was elected, and Robespierre obtained the result that the delegates of the National Constituent Assembly were not eligible for reelection. Despite the efforts of the Jacobins and the threats made by Marat, 264 "constitutional" right-wing delegates were elected, 345 nonattached members in the center and 136 left-wing Jacobins.

TERROR IN THE VOSGES

The assembly broke up on September 20th and I prepared to return home to the Vosges. I would join my wife and son in Saint-Dié, where we lived, and from now on I would follow the course of events from a distance. I would enjoy the calm of the Vosges after the turmoil of Paris.

Indeed, the situation in Paris was becoming more and more worrisome. The idea of the necessity of war was developing. It would have, for the revolutionaries, the double advantage of diverting the attention of the French people, who were very concerned about their increasingly precarious living conditions, and also of creating the upheaval that would overthrow the monarchy. War was thus declared on the King of Hungary and Bohemia where most of the nobles were exiled.

In May 1792, the assembly decreed that reactionary priests could be deported if requested by twenty citizens. Articles 4, 10, and 11 of the Declaration of the Rights of Man and of the Citizen were violated again! Louis XVI vetoed it. Urged by the Jacobins who were campaigning against this veto, the mob burst into the Tuileries Palace, ransacked it, and insulted the king. They were the sans-culottes, and constituted the armed wing of the Montagnards. For them, any riot was legitimate.

A manifesto by the Duke of Brunswick, summoning the French to submit without resistance or face reprisals by the Austro-Prussian army, ignited the powder keg. Louis XVI issued a denial of this to the assembly, but the damage was done. Robespierre called for the demise of the king, and the Tuileries were attacked on August 10th. The defenders, mainly Swiss Guards commanded by the Baron de Vioménil, resisted, but Louis XVI ordered the fighting to stop. The Swiss obeyed the orders, stopped, and were massacred. Despite all that had happened, the king still had not understood that his goodness was a weakness. In wanting to spare his attackers, he condemned his protectors and his family to torture. The king was imprisoned in the Temple tower and 3,000 persons deemed suspicious were imprisoned without cause. Article 7 of the Declaration of the Rights of Man and of the Citizen was again flouted.

During this attack on the Tuileries by the revolutionaries, Baron de Vioménil, who had remained close to the king and commanded the Swiss Guards, was wounded in the thigh. He managed to take refuge at the home of the ambassador of Venice, but gangrene set in and he died.[97]

---

[97] Personal letters of Baron de Vioménil. Introduction of the General Count de Grimoard.

The Terror was spreading. In August 1792 a decree forced 30,000 priests into exile, once again violating Article 10 of the Declaration of the Rights of Man and of the Citizen. In September, the sans-culottes entered the prisons in Paris and massacred the "suspects;" the police did not intervene. Elsewhere, a sham people's court was established. Articles 7, 8, 9, and 10 of the Declaration of the Rights of Man and of the Citizen were thus violated, without any objection from the ministers concerned or the leaders of the Revolution. Some had proposed to burn down the prisons in order to get rid of the prisoners, while others wanted to drown them.

The Legislative Assembly was replaced by the National Convention. The Jacobins had done everything to discourage the moderates, and the most extreme were elected. The first decision of the National Convention was to abolish the monarchy.

The sad fate of priests.

The French army began to win victories along our borders, and the conquered territories were systematically exploited, making it possible to bail out the sinking economy.

In December 1792, the National Convention decided to put Louis XVI on trial, the revolutionaries derisively calling him by the name "Louis Capet." At the end of a mock trial, the decision of which was predetermined, the king's death sentence passed with 361 votes in favor, including the vote of Philippe d'Orléans, and 360 votes against. The execution took place on January 21st, 1793. This king, who had been so miserable in his life, showed admirable courage in death. I was devastated at the thought that this execution had been approved by his cousin, but I had the satisfaction of learning a few months later that the traitor had also been guillotined.

Massacre of prisoners in September 1792. Anonymous engraving. BNF.

The death of the king provoked very strong reactions from foreign courts, especially since France had annexed the county of Nice and issued multiple warmongering statements. In early February, France declared war on England and Holland, who were joining Prussia and Austria. Soon after, Spain joined the coalition.

Military setbacks in France then mounted and soldiers deserted en masse. The National Convention decreed a levy of 300,000 men, which triggered revolts everywhere in France. These were put down everywhere except in the West, where a man named Jacques Cathelineau led the insurrection.

Famine was added to the military setbacks. The National Convention received a delegation from the forty-eight districts of Paris in February, who demanded measures to curb the famine. In response, a Gironde deputy retorted: "Who is talking about being hungry? The petitioners must be misled by the propaganda of the enemies of the revolution." Another, Tallien, added: "Let us stop talking about food from the rostrum. It can only have a very bad effect."

These gentlemen certainly had better things to do than to deal with the complaints of beggars! This did not prevent prices from soaring and the hungry from looting shops.

In March, the deputies of the National Convention voted to set up a revolutionary criminal court, which began to operate the scaffold in early April. The infernal machine was launched, and victims had to be found. To better create repression, the Committee of General Security and then the Committee of General Defense were created, which then transformed into the Committee of Public Safety.

At this time, dissension arose within the National Convention. The opposition between the Girondins and Montagnards became more and more virulent. The struggle ended with the arrest of twenty-nine Girondist deputies and the elimination of their group in the convention. It was then that Dumouriez, true to his opportunistic ways, began to change his attitude towards the Revolution. In April, he handed over to the Austrians the Minister of War and the four members of the National Convention who had come to arrest him, and went into their camp. The revolutionaries had guillotined many generals who had lost battles, so Dumouriez could expect the worst.[98]

This defection was to have far-reaching consequences on my life. In the absence of the real culprit, the revolutionaries needed to offer victims to the crowd to hide their incompetence. The representatives of the people of Saint-Dié, already worried, panicked when they heard the news of Dumouriez's betrayal. It was decided to disarm the so-called "suspicious" people, mainly the nobles and the clergy, but this operation was ineffective. More was needed. The Committee of General Security was therefore created. The General Council authorized it to arrest those who disturbed public order, to take communication of letters from abroad, and to receive and follow up on denunciations.

At that time, I was not worried, because the mayor, Jean Nicolas Barette, and the prosecutor, Jean-Baptiste Antoine, were of moderate opinion, even if they were forced, sometimes reluctantly, to follow events and obey the injunctions of the General Council. Nevertheless, on April 19th, at 7:00 in the evening, the police sergeant came to my house, seized the papers he found on my desk, and asked me to follow him to the General Council. I was not alone, as my two cousins, Sir Hugo of Spitzemberg, and Dominique de Bazelaire, were also arrested. We found

[98] Thirteen generals of the army of the Revolution were guillotined in 1793, and 24 in 1794.

ourselves with Philippe de Rozières, Canon of Thumery, Jean-Louis de Roûot, Esprit-Louis Geoffroy, former Canon Pierre-Charles Abram, Michel-Laurent Larminach, and Joseph Christophe. I am writing the names as they are, but at the time, the titles, particles, and family names were omitted. We were Hugo, Bazelaire, and Thibault!

We were read the decision of the General Council, the usual grandiloquent gibberish, which began thus: "The General Council, considering that the traitors from within have harmed the Republic a thousand times more than the slaves of the despots united against it..." Among other things, it included the words "crime against humanity" and "barbarous adversaries," in the midst of which was mentioned "a denunciation, to the truth devoid of evidence up to the present...." It was on the basis of this alleged denunciation that we were arrested, but mainly because of our kinship with emigrants and our noble origins. The decision foresaw the seizure of the papers of Hugo de Spitzemberg, Esprit-Louis Geoffroy, Dominique de Rozières, and myself.

The examination of these documents revealed nothing contrary to the Revolution, but the anxiety and agitation were such that the authorities had to explain that these harsh measures concerned people who were far from being unsympathetic to their countrymen. Only my cousin Hugo of Spitzemberg had made himself unpopular by his haughty behavior and by the fact that two of his sons, emigrants, were serving in the Army of the Princes.

The General Council posted a notice on the morning of April 20th, stating that our arrest was a mere precautionary measure and that no machination had been discovered. It was ordered that "their people, their wives, their children, their property be respected," which clearly meant that they feared bouts of violence.

We wrote a petition to the administrators of the Vosges, recounting the conditions in which the arrest had taken place, stating that none of us had put up any resistance and that the papers seized and examined showed no evidence against the persons concerned. We therefore asked for our release, at least temporarily, by proposing to post bail. The town hastened to address this petition to the departmental administration, supporting it as best it could by declaring that it had not found anything against us. It asked that we simply be consigned to our homes. The District Decree of the department taken on May 16th was immediately communicated to us. It rejected the request of eight of us and let the General Council decide the fate of Larminach and Christophe. Finally, all detentions were maintained, and that of Dr Charles Aubry was added. Once again, the Declaration of the Rights of Man and of the Citizen was violated. We were deprived of liberty without cause and without trial.

We were transferred to the building of the old bishopric on June 27th. On July 1st, the council authorized visits from close family members, granted us a three-hour walk a day, and two servants at our disposal for meals. Then began a detention with no defined limits.

The situation was getting worse across the country. Several cities revolted to protest against the elimination of Girondins: Marseille, Avignon, Bordeaux, Lyon, and the Vendée, where the Republican forces were held in check. It was the same at the borders. In Paris, the Montagnards, now rid of the Girondins thanks to the sans-culottes, had to resist the pressure of the latter as well as that of the angry people. The pressure was even stronger because the food shortage was getting worse. The convention had to yield and decreed the mass levy demanded by the sans-culottes.

In Saint-Dié, contingents from neighboring districts were formed. Many of the young recruits spoke only of making prisoners disappear so as not to leave enemies behind, as had been done in Paris the year before. On the evening of Friday August 30th, a guard of about fifteen citizens came to spend the night at the bishopric. They warned us that they had heard horrific threats of massacre, and on Saturday we learned of more threats and insults. We trusted the municipal officers though, and we considered ourselves safe. However, the threats were renewed with fury in the night from Saturday to Sunday, and on Sunday morning, the young recruits about to depart had regrouped with their families. The inns were full; the wine flowed freely; and about half-past 1:00, the crowd demanded that the Directory and the administrators leave with them. The mayor ordered the commander of the National Guard to restore calm. The crowd then moved towards the bishopric.

Around 3:00 a.m., we were assailed by a crowd of drunken people, armed with spikes and axes, shouting "Give us the prisoners!" We fled just as they entered the bishopric. Philippe de Rozières had difficulty running because he suffered from gout and hid in a bush where, luckily, the pursuers did not see him. He witnessed the arrest of Hugo of Spitzemberg and the beginning of his horrific martyrdom. Philippe de Rozières was found by two citizens, who helped him escape. As for me, I managed to escape the pursuers, like my other companions did.

I later learned that the bishopric had been completely looted, but much worse, that Hugo had been taken and, despite the efforts of the district administrators, the prosecutor, and the members of the General Council, carried away by the mob. His ordeal was long: he was walked through the streets under blows and insults, and taken to a chapel outside the city. There, his executioners mutilated him and then, finally exhausted, ended up killing him. It was 6:00 in the evening. They brought Hugo's corpse to his home and destroyed or stole everything in his house.[99]

The next day, the looters went to Hugo's house again to finish looting and drink what was left of the wine. They then went to my house where the looting was more organized, and were even joined by members of the National Guard. Other homes suffered a similar fate. Everyone knew who the perpetrators of these crimes were, yet they were never punished. The convention, having been informed of these facts, suspended the proceedings. Aside from Philippe de Rozières, we were all safe.

The representative of the people, Guyardin, arrived from Épinal and issued a decree directing us to become prisoners again. I had no choice, as staying hidden was impossible and sooner or later I would have been reported, and exile would have endangered my wife and son. So I decided to go to Épinal. I found Philippe de Rozières, along with Martimprey de Choisimont,[100] Dominique de Bazelaire, Esprit-Louis Geoffroy, Abbé Larminach, and Dr Aubry. As for Christophe Roüot, the two canons of Thumery and Pierre-Charles Abram, they had gone abroad. We were imprisoned in the old building of the Annonciades. For how long? We had no idea.

Things were even worse in Paris. The Committee of General Security was renewed and relied on such a broad definition of suspects that anyone could be arrested. After a mock trial, Marie-Antoinette was guillotined. She showed exemplary courage. Then, it was the turn of the Girondins. Several of them who were already on the run committed suicide to escape their fate. From October, the Revolutionary Tribunal met daily, sending each day those condemned to death to the "national razor," many for trivialities, a word too many, or just one unfortunate statement.

Many cities revolted against the National Convention, but repression was severe, especially in Bordeaux, Marseilles, and Lyon, where Fouché ruled. He would later become Minister of the Interior. The worst took place in the Vendée. Westermann's statement to the convention was repeated: "I crushed the children

---

[99] Bouvier, F. op. cit.

[100] François-Louis's wife was of the family Martimprey de Choisimont; he was probably his brother-in-law.

under the hooves of the horses and massacred the women. I don't have a prisoner to reproach myself with. I exterminated everyone." It was nothing but killings, rapes, pregnant women disemboweled, boats of prisoners sunk. The Vendéens died by the thousands. Human nature revealed its worst through the Westermanns and the Fouchés.

A vast movement of dechristianization was launched at this time. Saints were suppressed, as well as religious festivals. Christmas became the day of the dog! Churches were looted and closed. Anti-religious demonstrations were organized and the statues of the saints destroyed. Joseph Fouché, who had already raged in Lyon and Nièvre, was one of the fiercest.

Three factions began to tear each other apart in Paris. Robespierre launched a wave of repression against conspiracy and traitors to assert his power. The revolutionary government then eliminated the extreme left by sending the Hebertists to the scaffold; then it was the Dantonists' turn. Despite the increase in detention facilities in Paris, the prisons were full. The security committee said it was purging them, and invented conspiracies or rebellions. Moderatism became a crime punishable by death. Among the guillotined were 73-year-old Malesherbes and his family, generals who had suffered defeat, and fermiers généraux, financiers who managed the king's revenue, such as taxes. I thought we had reached the height of persecution, but it was not at all! In June 1794, a decree further accelerated the procedure of the revolutionary tribunal: no more questioning before trial, only prosecution witnesses accepted, no more lawyers…

However, there were many members of the convention who disapproved of these methods. Some spoke out, but Robespierre threatened to ask for the heads of the schemers, so the convention complied, not without resentment. The guillotine became more active than ever in July 1794. Among the convicts were nobles, priests, but also nuns, servants, shopkeepers… I learned that as a result of the June decree, nearly 1,500 people were guillotined in a month and a half.

Meanwhile, the starvation of the poorest was increasing and taxes were not coming in. The republic ordered the armies to plunder the conquered countries. The instructions were clear: "Wrest from the enemies all of their resources, all their means of existence."

The only point of satisfaction in this world of horrors was that slavery was abolished in February 1794. When I was a member of the National Constituent Assembly, we succeeded in achieving freedom and equality for free Blacks and mulattoes born to free fathers and mothers. This victory would not last, however, because Bonaparte would re-establish slavery in 1802.

Then astonishing news reached us at the end of July: Robespierre had been arrested. The Montagnards had decided to eliminate the tyrant as they feared more and more for their lives. Robespierre was guillotined along with a hundred of his friends on July 28th, an unforgettable date. All over France, this news was greeted with joy. We were beginning to hope for an end to our detention. We learned that many suspects had been released earlier in the month. We sent a joint letter to the authorities on August 17th asking for our release. Despite the arrival of the representative charged with ending the Terror, our situation was only examined later. It was not until November 11th that Philippe de Rozières, who was suffering terribly from gout, was released under certain conditions. A few days later, we were also allowed to return home.

My wife and I decided to retire to Blâmont, where she owned some properties. I was enjoying my newfound freedom, hoping that this troubled period would soon be over.

Nothing really changed in the Vosges, which had gone through these years in a quieter way than many other regions. Apart from the horrible martyrdom of my cousin Hugo, there were few deaths. The criminal court of the Vosges had pronounced nine death sentences, all against priests and their servants. Eleven people were brought before the revolutionary tribunal in Paris and beheaded. On the other hand, there was no counting of the multiple lootings and pillages. The forests had been ravaged, both to meet the demands of the authorities and by individuals who had taken advantage of the powerlessness of the agents of the forest service.

The passage of troops and convoys had damaged and gullied the roads that were no longer maintained. Famine was also present; many villages had only bad bread, and there was not enough for everyone.

The situation in Paris remained confused. The Thermidorians, conspirators who eliminated Robespierre, did not know how to get out of the Terror. They decided to suppress the Jacobin Club and the popular societies, and freedom of worship was restored in February 1795. They still had to confront the sans-culottes, who demonstrated and demanded bread. To bring them back to order, the districts of Paris were cleared of them and thirty-six people were sentenced to death, including several extremist delegates.

For its part, the royalist movement regained its colors, but the convention prevented their return by requiring that two-thirds of delegates come from the convention. This decision provoked an insurrection, and twenty-five thousand people marched to the convention to demand that equal rights be respected.

Bonaparte fired the cannons with machine guns, causing 300 deaths. Those petitioners who were unarmed dispersed.

The Directory tried to protect itself both from the unrest of the hungry people and from the monarchists who were increasing in number, so much so that the Council of Five Hundred became a royalist majority from spring 1797.

Outside France, the war continued. General Bonaparte, appointed commander-in-chief of the army of Italy, accumulated successes and the conquered countries were forced to contribute in cash, in kind and in works of art. Paris demanded that funds be sent and that works of art be looted. Without this sacking, France would have been bankrupt. Bonaparte is said to have brought more than 50 million from Italy to France. The general was a true hero admired by all, so much so that the Directory sent him to conquer Egypt to supposedly cut off the English route to India; they were very happy to send off a man who was becoming too popular.

The Egyptian campaign, a success at first, ended in disaster. The fleet was destroyed at Aboukir by the British. Bonaparte, informed of the problems of the Directory, left for Paris, leaving behind an army that was doomed to be defeated. Fortunately, they were able to surrender to the English, who repatriated the survivors.

France was still going through a serious financial crisis. The army was suffering setbacks and the political climate was uncertain; the end of the regime seemed increasingly inevitable. On November 10th, 1799, after many experiences and an operation by the army that caused no casualties, Bonaparte was appointed consul, accompanied by Sieyès and Ducos. I followed from afar the rise of this man who had become French thanks a little to me, since I had participated in the conquest of Corsica! He began by dominating the other two consuls and then the entire political life of the country. He established a dictatorship for his benefit, which some called for to put an end to anarchy. In any event, no one was in a position to challenge him.

Plato was right once again, when he wrote that democracy inevitably degenerates and is replaced by a tyrant. Bonaparte had now become Emperor Napoleon, and continued his conquests. The number of both civilian and military deaths was counted in the millions, but I was grateful that he had brought order and prosperity back to France.

As for myself, I became a municipal councilor of Mirecourt, president of the assembly of the canton of Baccarat in 1806 and a member of the General Council of Meurthe. We were now living in Deneuvre, in the house of the provost of the Collegiate Church.[101]

In my opinion, Bonaparte committed three errors which led to his loss. The first was the assassination of Louis Antoine de Bourbon, Duke of Enghien, which earned him an even stronger hatred of all the royal courts of Europe. The second was to try to conquer Spain, and the third was to try to conquer Russia. The man was a victim of his ambition and military genius. Napoleon abdicated in April 1814 and Louis XVIII, brother of Louis XVI, became King of France. Napoleon returned to France in 1815, only to suffer a final defeat that earned us even harsher treatment than that inflicted on us in the previous treaty of Vienna.

Our country was ruined humanly, politically, and economically. We had lost all our colonies. Our influence in the world was reduced to nothing. Our territory was smaller than at the accession of Louis XVI. In addition, we had to pay a heavy war indemnity for the maintenance of foreign troops stationed on our soil. After twenty-six years of misery, we again found the Bourbons at the head of France, within the framework of a constitutional monarchy.

In 1812, my wife Charlotte died. I am now alone, but I have the satisfaction of being able to complete these memoirs when our country has regained its serenity.

François-Louis died on December 5, 1816
at Deneuvre (Meurthe)

[101] Feve, ML. *Dernier Voué de Baccarat: François-Louis Thibault de Ménonville.*

JOURNAL OF THE SIEGE OF YORKTOWN
By Major General Aide François-Louis Thibault de Ménonville

<u>*October 6–7*</u>

Opening of the trench;
Major General, baron de Vioménil;
Brigadier, Count Custine;
Bourbonnais, 2 battalions;
Soissonnais, 2 battalions;
Night workers, 1,000 men.

The trench is opened by a parallel to the right of which is a ravine adjoining the Pigeon Hill redoubt. The right joins the left of the Americans one, cutting at right angles the road from York to Hampton.

This parallel is defended by four redoubts, two on the American side and two on the French side.

The work of the Americans, which in fact is one with ours, stops at the river. Their activity during the night consisted in building their part of the parallel.

At the same time, a trench was opened on our left, joining the upper part of the river and defended by a battalion of the Touraine regiment, its grenadiers and light infantry. A battery was started in order to reach enemy ships upriver.

The enemy discovered this work very early. It attracted their attention so much that they ignored the main part, on which they did not fire a single shot. As on previous nights, they contented themselves with firing at the redoubts which they had abandoned to us, and those built by the Americans on both sides of the Hampton road, and which lie behind our work. Their shots were inaccurate and only slightly bruised an officer of the *Royal Deux Ponts* and a slightly more serious one on a soldier in his regiment. Both were with the workers.

An artillery officer In the Touraine regiment was seriously wounded as well as six grenadiers, two lightly, and a soldier of the Agenois regiment.

At daybreak, the works of the main assault were ready almost everywhere to receive their troops.

The day was spent improving the parallel with 400 men taken from the battalions of the trenches.

October 7–8

Major General, Chevalier de Chastellux;
Agénois, 2 battalions;
Saintonge, 2 battalions;
Night workers, 900 men.

Tonight, 500 workers are employed, under the direction of the engineers, to begin the communication at the rear and right of the parallel, improving it and the redoubt, and making a communication trench with the batteries.

The other 400 night workers are employed with those of the artillery in constructing the batteries.

On the part near the head of the river, the construction of the battery continued. It was ready to open fire at daybreak.

400 day workers were employed in the trenches to perfect the work of the previous two nights and to continue the construction of the batteries.

The construction of batteries also began in the upper part of the work of the Americans.

Injured on the main construction site: 6.

The arming of the batteries started on the night of the 7th to the 8th.

Americans - On the right, adjacent to the river, a battery of 6 guns and 4 mortars.

Near redoubt 5 of the French, a battery of 5 guns.

French - n° I. 1 large battery composed of 4 16-pounders; 2 12-inch mortars; 4 8-inch mortars; 2 8-inch howitzers; 4 16-pounders.

A little behind the parallel and a little to the right of Hampton road.

No. II. A battery of 4 24-pounders, also behind the parallel and to the right of the ravine to which it ends.

No. III. One of the 3 24-pounders, in the direction and rear of the entrance of the parallel.

No. IV. One of the 3 24-pounders to the left of the ravine where the parallel stops.

October 8–9

Major General, Marquis de Saint-Simon;

Brigadier, Count Custine;
Gatinois, 2 battalions;
Royal Deux-Ponts, 2 battalions;
Auxiliaries, the grenadiers of Soissonnais and Saintonge;
Night workers, 800 men.

Half of the night workers were employed under the direction of the engineers to finish the communications begun the previous nights. The others, along with those of the artillery, to continue the construction of the batteries and to begin a new one, No. V. in front of the parallel to the left.

400 workers were employed in the same work during the day near the French troops.

The Americans continued the work of the previous days and nights.

Their battery of 6 guns and 4 mortars along the river was ready to open fire two hours before nightfall.

The one upriver began firing at an enemy frigate about three o'clock, forcing it to break its moorings and withdraw.

Killed 1. Injured, 1.

October 9–10

Major General, Count Viomenil;
Bourbonnais, 2 battalions;
Soissonnais, 2 battalions;
Auxiliaries, the light infantry of the Agénois and Gatinois;
Night workers, 700 men.

400 workers were employed during the night under the direction of the engineers to protect the redoubts of the parallel and to improve communications. 300 workers along with those of the artillery to continue the batteries.

The following batteries opened fire during the day:

American batteries, 5 guns;

French batteries, No. I and No. II. Two hours later, enemy fire ceased completely.

200 day workers were employed near the French battalions to improve communications and continue unfinished work on the batteries. Wounded on the main structure, 2 men.

October 10–11

Major General, Baron de Vioménil;
Brigadier, Count Custine;
Agénois, 2 battalions;
Saintonge, 2 battalions;
Auxiliaries, the light infantry of Soissonnais and Royal Deux Ponts;
Night workers, 300 men.

Night workers were employed to improve batteries and redoubts during the night. The fire continued with bombs from the American batteries and our battery No. I.

Red-hot cannonballs were fired during the attack on the ships. The *Charon* and two transports were set on fire.

Our batteries No. 3 and 4 continued firing during the day.

The enemy began firing around three o'clock in the afternoon. A few shots from the right hit Battery No. I and nearby structures.

Killed, 1. Wounded, 3.

October 11–12

Major General, Chevalier de Chastellux;
Gatinois, 2 battalions;
Royal Deux Ponts, 2 battalions;
Auxiliaries, the light infantry of the Bourbonnais and the grenadiers of Saintonge; Night workers, 800 men.

Under the direction of the engineers, 750 night workers began a second parallel about 400 feet in front of the first. The end was located at the large ravine where the first one ended, which also served as access.

The American workers built their part of the parallel toward the right, where it extends to the other side of their five-gun battery. An access starting at the ravine to the left of our redoubt joins it on the right, and another begins to the left of our battery n° 2 and proceeds in zigzag to the second parallel. This parallel is protected by two redoubts.

50 night workers were employed to finish battery No. 2, and to repair the others.

In order to protect the works and hide them from the enemy, our mortars and howitzers, as well as those of the Americans, continued firing throughout the night. Our gun batteries maintained a moderate fire. The enemy sent bombs and fired a few cannon shots, but did not interrupt the work that was completed at daybreak. That was also the case among the Americans.

During the day, our Battery No. V opened fire and our batteries continued to fire until nine o'clock, when, the fire was suspended fearing it would disturb the American day workers. When the enemy reopened fire, our batteries Nos. 3 and 4 were put back into action.

300 day workers drawn from the French regiments continued work on the parallel.

Wounded on the main structure, 4 men; and the Touraine, 3 men.

October 12–13

Major General, Marquis de Saint-Simon.
Brigadier, Count Custine.
Bourbonnais, 2 battalions.
Soissonnais, 2 battalions.
Auxiliaries, grenadier of Agenois and Gatinois.
Night workers, 600 men.

Night workers completed most of the previous night's work; 300 day workers completed the parallel and started building the batteries.

Killed 6; wounded 11, in the grand attack. Messrs. de Miolis and Dursu, a Soissonnais officer wounded.

October 13–14

Major General, Count de Vioménil;

Agénois, 2 battalions;
Saintonge, 2 battalions;
Auxiliaries, grenadiers of the Soissonnais and Royal Deux Ponts;
Night workers, 600 men.

300 night workers were employed to improve the redoubts and the rest worked in the trenches. This work was continued by 300 day workers from the French battalions.

Killed, 1; wounded, 28, in the grand attack.

The arming of the batteries began on the night of the 12th to the 13th.

No. VI. Between the two communications and a little ahead of the second parallel, 6 guns.

No. VII. To the left of the communication trench on the left of the parallel, 6 guns.

No. VIII. In the parallel to the right of the left redoubt, 6 guns.

No. IX. In front of the parallel, 8 mortars, 2 howitzers.

October 14–15

Major General, Baron de Vioménil;
Brigadier, Count Custine;
Gatinois, 2 battalions;
Deux Ponts, 2 battalions;
Auxiliaries, the grenadiers of Saintonge, the light infantry of Bourbonnais, Agénois and Soissonnais;
Night workers, 800 men.

The order was given to attack the enemy's two advanced redoubts, the first along the river and the other to its left. This was done after dark. The American light infantry, supported by French trench battalions, all commanded by Lafayette, attacked the redoubt near the river and finished with the bayonet with 4 officers wounded and about 20 men killed or wounded.

The French troops charged with attacking the other redoubt departed on the right side of the American 5-gun battery in the following order:

The companies of grenadiers and light infantry of the French regiments commanded by Count Guillaume de Deux Ponts, second colonel of the regiment bearing his name, and of lieutenant colonel de l'Estrade, of the Gatinois; the first battalion of Gatinois regiment; the auxiliary grenadiers and light infantry from the trenches, with the exception of the Soissonnais light infantry were ordered to make a feint on the left of our main assault. This division was in support under the orders of colonel Marquis de Rostaing, of the Gatinois regiment.

French Major General Baron de Vioménil, commanding the whole, headed towards the redoubt with his troops in perfect order and silence. The enemy quickly identified the column and opened a strong fire of musketry. The palisades were in better condition than expected after several days of heavy artillery bombardment. Despite heavy enemy fire, the pioneers of the regiment opened passages through which the grenadiers of Gatinois and Deux Ponts entered the trench and at the same time as the pioneers who were forced to cut down palisades to open the chevaux-de-frise of the redoubt. These grenadiers took advantage of these openings to climb the parapet where they regrouped in such a way that the enemy could do nothing but surrender.

We took 40 officers and 3 soldiers and counted 18 dead. The rest of the 170 occupants fled.

Our losses in this attack were about 80 killed and wounded.

The enemy began a heavy fire of cannonballs and mortar shells on the redoubt we had taken, killing and wounding a large number of soldiers. As soon as we captured the redoubt, 500 workers arrived on the right of the second parallel to extend it to the redoubt. The American workers extended this parallel between the two redoubts, opened an entrance from the first which opens between their large battery and their first redoubt on the right, and continued to the enemy redoubt they took. All this work was done with the greatest speed and was very advanced at daybreak.

The feint on the left of our work was carried out a little too neatly. We lost a few men.

At nightfall, the enemy's attention was drawn to the head of the river by a diversionary maneuver executed without loss by the Touraine regiment.

October 15–16

Major General, Chevalier de Chastellux;
Bourbonnais until the evening of the 15, 2 battalions;
Agenois auxiliary for the night;
Soissonnais, 2 battalions;
Auxiliaries, light infantry of the Royal Deux Ponts;
Night workers, 500 men.

100 night workers were employed to improve the batteries. The other 400 to improve the parallel and the redoubts. About five o'clock in the morning the enemy made a sortie. They infiltrated a redoubt and our batteries where they spiked four cannons that could fire again six hours later. Battery No. VI opened fire about noon.

Killed 1; wounded 37 during the main assault. Also wounded: Mr. de Marin, Captain of the Soissonnais regiment, lieutenant de Bargues of the Bourbonnais, Lieutenant of Artillery de Pusignan; Captain de Bourgimont, of the Agénois, taken prisoner.

October 17–18

Major General, Count de Viomenil;

Bourbonnais, 2 battalions;
Royal Deux Ponts, 2 battalions.

A second flag came about three o'clock with proposals that allowed the fighting on both sides to stop, until the signing of the capitulation on the 19th at noon.

Killed 2; wounded, 1.

Monsieur de Bellanger, lieutenant of the Royal Artillery Corps, killed.

Monsieur Drouillet, lieutenant of the grenadiers of the regiment of Agenois, was killed on September 1 in a diversion made on the redoubt on the right of the enemy.

RECAPITULATION OF KILLED AND WOUNDED.

DATE	KILLED Soldiers	KILLED Officers	WOUNDED Soldiers	WOUNDED Officers
6....	.....	.....	8	2
7....	.....	.....	6	......
8 ...	1	.....	1	......
9....	1	.....	2	......
11....	.....	.....	3	......
12....	6	.....	7	......
13	1	.....	11	2[a]
14....	46	.....	62	6[b]
15....	1	.....	37	5[c]
16....	1	.....	10	1[d]
17....	2	.....	1	1[e]
Sept				1[f]
Total	59	1	176	17

Grand total, 253 men killed and wounded.

The Magazine of American History, October 1881.

BIBLIOGRAPHY

Balch, T. *Les Français en Amérique pendant la guerre de l'Indépendance des États-Unis: 1777–1783.*

Bardy, H. *Les otages de Saint-Dié.* Le Pays Lorrain.

Benoit, A. *L'École des Cadets-Gentilshommes du roi de Pologne* à *Lunéville. 1738–1766.*

Berthier, L-A. *Journal.*

Blanchard, C. *Guerre d'Amérique. 1780–1783. Journal de campagne de Claude Blanchard.*

Bodinier, G. *Les officiers de l'armée royale.*

Bodinier, G. *Dictionnaire des officiers généraux de l'armée Royale 1763–1792.*

Boulesteix, P. *L'École royale du génie de Mézières et sa belle descendance (1748–1794).*

Bouvier, F. *Les Vosges pendant la Révolution.*

Brette, A. *Recueil de documents relatifs à la convocation des* É*tats généraux de 1789.* Tome 2

Bryant, C. *The Siege and Surrender of Yorktown. Van Blarenberghe's Paintings and the Metz Regiment of Artillery.*

Burke, E. *Réflexions sur la Révolution en France.* Les Belles Lettres.

Burke, E. *Lettre* à *un membre de l'Assemblée nationale.* Mille et une Nuits.

Carrington, D. *La révolution Corse (1729–1769).* Vicomte Chateaubriand, F-R. *Mémoires d'Outre-Tombe.*

Chotteau, L. *Les Français en Amérique, la guerre de l'indépendance (1775–1783).*

DEPUTY OF THE NOBILITY TO THE ESTATES GENERAL

Comte de Clermont-Crèvecœur, J-F. *Journal de la guerre en Amérique pendant les années 1780, 1781, 1782, 1783.*

Marquis de Chastellux. *Voyage dans l'Amérique septentrionale 1780, 1781, 1782.*

Baron Cromot du Bourg, M-F-J. *Journal d'un officier français, 1781.*

Dumouriez, C-F. (1848). *Mémoires du général Dumouriez.* Librairie Firmin Didot.

Dupont d'Aubevoy de Lauberdières, L-F-B. Journal de l'armée de Rochambeau, 1780–1783.

Fève, ML. *Dernier voué de Baccarat: François-Louis Thibault de Ménonville.*

Gardiner, A. *The society of Cincinnati.*

Hindie Lemay, E. *La vie quotidienne des députés aux États Généraux.*

Hindie Lemay, E. *Dictionnaire des Constituants.*

Lallement, L. *L'expédition de 1769 en Corse et son annexion à la France. Info Bretagne.com.*

Lallement, L. (1889). L'expédition de Corse de 1769. Bulletin de la Société Polymathique du Morbihan, 3e Semestre.

Magazine of American history, 1780, Vol IV, October 1781. The French in Newport, *Newport history*, n° 249–250.

Pautet, S. (2016). *Produire une élite savante et technicienne* à *l'École du Génie de Mézières.* Artefact 4.

Quetel, C. *Crois ou meurs.* Taillandier.

Abbé Robin, *Nouveau voyage dans l'Amérique septentrionale en l'année 1781 et campagne de l'armée de M. le Comte de Rochambeau.*

Robinet, A. (1949). *L'École royale du génie de Mézières* (1748–1794). *Revue d'histoire des sciences et de leurs applications*, tome 2, n° 3.

Comte de Rochambeau, J-B. D. *Mémoires militaires, historiques et politiques.*

Stone, E. M. *Our French allies.*

Thibaut de Ménonville, F-L. (1881). Journal of the siege of Yorktown. Magazine of American History.

Thibaut de Ménonville, F-L. *Lettre* à *Edmund Burke.*

Thibaut de Ménonville, F-L. *Journal du siège de York.* Manuscript formerly held by the de Chastellux family.

Trintignan, J. de. *La France au secours de l'Amérique.* Éditions SPM.

Verger, J-B-A. de. *Journal des évènements les plus importants.*

Baron du Houx de Vioménil, A-C. *Lettres particulières.* Warning of the General Count of Grimoard.

DOCUMENTS

Letter addressed to François-Louis de Ménonville, confirmation of nobility, signed by Louis XV and dated 20 August 1733.

Letter addressed to François-Louis de Ménonville, signed by Monteynard and Vioménil, dated 10 October 1772.

Archives of Vincennes, service records.

Letter from the Comte de Rochambeau to François Louis Thibaut de Ménonville from 21 September 1788.

National Assembly, parliamentary archives.

La Jaune et la Rouge, magazine of the Ecole Polytechnique, March 2020.

Annales historiques de la Révolution française, 27th year, number 140.

Made in the USA
Monee, IL
04 November 2026

67379261R00066